JONATHON PORRITT

Seeing Green

The Politics of Ecology Explained

BASIL BLACKWELL

© Jonathon Porritt 1984

First published 1984

Reprinted 1984, 1985, 1986 (twice)

Basil Blackwell Ltd
108 Cowley Road, Oxford OX4 1JF, UK

Basil Blackwell Inc.
432 Park Avenue South, Suite 1503,
New York, NY 10016, USA

British Library Cataloguing in Publication Data
Porritt, Jonathon
  Seeing green.
  1. Conservation of natural resources—Political aspects
  I. Title
  333.7'2   5944

ISBN 0-631-13892-7
ISBN 0-631-13893-5 Pbk

Also included in the Library of Congress Cataloging in Publication lists.

Typeset by Stephen Austin/Hertford England
Printed and Bound in Great Britain by
T. J. Press (Padstow) Ltd, Padstow, Cornwall

When the forms of an old culture are dying, the new culture is created by a few people who are not afraid to be insecure.

Rudolf Bahro

Contents

Foreword

The history of green politics is still a relatively short one, and there are many for whom this is still a relatively new area of concern. So it was good news when I heard that Jonathon Porritt, a close friend of mine and a respected and committed member of the international green movement, was writing a book about green politics. I knew that he would be able to describe best what is happening in the UK, and to set it in an appropriate international framework. But *Seeing Green* is not just another analysis of the politics of ecology: it is a very personal book written by someone who has experienced and been enriched by the green perspective at every level of his life. And I am particularly pleased to have been asked to write the foreword to it, for I share many of the views and evaluations which Jonathon puts forward.

He describes green politics as the rediscovery of old wisdom made relevant in a very different age. In an age of Orwellian reality, in an age of nuclear terrorism when Governments hold their own men, women and children as hostages, in an age when the European Economic Community has been co-responsible for world hunger through its insane agricultural policy and its policy of 'institutionalized waste', in an age when economics, GNP, the profits of multinational companies and worldwide competition continue to take precedence over all the most basic human needs, which remain unfulfilled, and in an age when mass destruction is happening all around us all the time as 15 million children die every year in the Third World from malnutrition and disease – how can one fail to agree with Jonathon when he writes: 'Opposition to this dominant world view cannot possibly be articulated through any of the major parties, for they and their ideologies are part of the problem. Green politics challenges the integrity of those ideologies, questions the philosophy that underlies them, and fundamentally disputes the generally accepted notions of rationality'? When humanity shrinks from recognizing that it is in the process of destroying itself, we have no option but to accept that the present political system is entirely bankrupt.

The history of die Grünen in the Federal Republic of Germany, which I have experienced from its very beginnings, is very different from that of the Green Party in the UK. But these two sister parties share a profound commitment to the same sort of radical green politics based on non-violence in all areas of life. It is, of course, very difficult to build up a democratic grass-roots movement of this sort. We are all people of the old world trying to create a new one. We must constantly be asking ourselves whether we now want to continue to support the status quo, seeking to cope with crises *after* the event by using outdated and ineffectual methods of crisis management, or whether we should be aiming for a completely different kind of politics and conflict resolution. It becomes increasingly important to commit oneself to what one feels to be right, to refuse to be forced into choosing the so-called 'lesser' evil. We often refer to die Grünen as an 'anti-parties party', by which we mean a party always capable of deciding between morality and power, a party which will be prepared to counter repression with creative disobedience, a party capable of combining bold imagination with efficient work and of grasping the connection between peace in the world and peace in each individual.

One of the most important chapters in this book is the one concerning Green Peace (chapter 11). As Jonathon says: 'Lasting peace can be based only on a genuine understanding of the relationships between people and planet.' Peace movements and green movements all over the world should be actively developing non-military alternatives to the present defence policies, based on 'social defence' or 'civil resistance'. You cannot do away with violence by using violence. We should not be the ones to have to think about military alternatives or short-term objectives in a world of anti-aircraft defences, fighter aircraft and high-technology weapons. We must be the ones who present a *real* alternative to the dead-end self-destruction of the present arms race. As we have written in our Peace Manifesto: 'Social defence is defence by non-military means against a military attack from inside or from without, and is based on the general idea that a society cannot be controlled if it is not prepared to co-operate with the oppressor.' Peaceful and non-violent methods of resistance in the tradition of Mahatma Gandhi, Martin Luther King and Dorothy Day must be developed and practised at a local as well as at an international level.

Green movements and green parties have a crucial responsibility here: 'Not the power of the "powerful" of this Earth, not the cunning of politicians, not the cold-blooded attitudes of the strategists will achieve peace. No, truly "disarming" are the warmth, the hope, and the courage of millions, who individually "powerless", are, together, irresistible.' Those were the concluding words of our Peace Manifesto in 1983. At the time of my writing this foreword, we are already surrounded by Pershing II in Mutlangen, Cruise missiles at Greenham Common and a massive proliferation of every kind of nuclear weapon. Our Governments, whether they are led by Margaret Thatcher or by Helmut Kohl, give little ground for hope. They, along with Ronald Reagan and Mikhail Gorbachev, rely on their secret services and on the daily increase of the weapon arsenals. They incite fears and conjure enemies in the midst of their populations.

Yet we cannot give up hope. Martin Luther King found an answer which we must listen to if our green movement is to succeed:

> We are now faced with the fact that tomorrow is today. There is such a thing as being too late. Over the bleached bones of numerous civilizations are written the pathetic words: 'Too late.' If we do not act, we shall surely be dragged down the dark corridors of time reserved for those who possess power without compassion, might without morality and strength without sight.

It is my hope that this inspiring book, with its clear message that it is *not* too late, will encourage many to act now by embracing the politics of life.

Petra Kelly
June 1984

Preface

On 12 April 1984 the *Guardian* carried an article with the arresting headline: 'British "Greens" worry PM'. About time too, I thought to myself. For there seemed no reason why Mrs Thatcher should be immune to the same symptoms of anxiety already experienced by conventional politicians the length and breadth of Europe.

Their anxiety is the product of *our* optimism – the optimism of all those involved in the burgeoning green alternative. For ten years or so, conventional politics has been on the defensive: its policies have failed, its ideals have been forgotten, its visions have faded. At the same time, amidst all that encircling gloom, there has grown a point of light to which more and more people are now turning. That's what this book is about: both the darkness and the light, both the grey and the green. The first part seeks to provide some explanation of what it means to be green; the second, of why we now find ourselves in such a mess; and the third, of how we might get out of it. The fourth part is by way of a challenge to those who have followed the argument to its conclusion and have realized that only through a genuinely green revolution is today's gloom likely to be dispelled.

It wouldn't always have been possible to start a book on green politics in so positive a vein. Since the early 1970s there have been some difficult times and more than our fair share of disappointments, false starts, prolonged troughs, and moments of doubt. It accepted notions of rationality.

still isn't easy; though we have all become adept at combining a
certain 'optimism of the will with pessimism of the intellect',
there are times when the balance is hard to maintain. Yet few of
the hundreds of thousands of people involved in green politics
today feel as tentative or as apologetic as they once did. There's no
need for people to remain 'closet ecologists', hiding their identity
and their aspirations, for that sense of fragile impermanence has
gone.

So too has that shade of blighted despair that once darkened the
utterances of early ecologists. This book is not just another
catalogue of eco-doom, another valedictory moan as we vanish
down the plughole of a blighted biosphere; I would like to think
I'm still in the business of saving planets, not burying them. Nor is
it just an extended party political manifesto, or just another book
about party politics, for if it were only that, then I would not have
done justice to the significance of the green perspective.

At a very personal level, it's an attempt to give voice to some of
the inspirations and frustrations of being green – and there are
plenty of both. The very business of being involved in 'minority
politics', while at the same time being part of the most dynamic
social and political movement since the birth of socialism, is just
about the most frustrating thing you can imagine! Persuading
people to vote for the inevitable is a remarkably tricky task. We
have all become familiar with the 'armadillo syndrome' – roll up
tight in a ball, and let the world go by. You may beat such people
around the head, tickle their tummies, tell them that their supply
of ants is going to run out in the next decade, tempt them with
visions of verdant armatopias, but they are impervious to it all.
They just want you to leave them alone to go about their armour-
plated business. There are still many who willingly or unwillingly
acquiesce in the system, pledging loyalty to the dominant world
view and its attendant myth of progress. And there's still a great
deal of hypocritical humbug, of narrow, linear thinking that
postures as politics.

Opposition to this dominant world view cannot possibly be
articulated through any of the major parties, for they and their
accompanying ideologies are part of the problem. Green politics
challenges the integrity of those ideologies, questions the philosophy that underlies them, and fundamentally disputes today's

For every armadillo there's someone out there who's already gone, or who's in the process of going, green. 'Going green' is always a rather personal thing; as someone who now spends his entire life involved in green politics, it amuses me to think how haphazardly I myself became interested. I first started out in life as a teacher, and back in 1975 or thereabouts, someone lent me an article about education in a magazine called the *Ecologist*. I read it, then read the rest of the magazine, and found it excellent. So I became a subscriber, being quite unable to find it in any newsagent. I soon began to notice the occasional scruffy little adverts exhorting me to join the Ecology Party. At that stage the idea of getting seriously involved in politics had never entered my mind. Nonetheless, I felt I had to find out more about this quaint organization called the 'Ecology Party'.

So I duly wrote in for some literature, only to be told, in the politest of letters, that the party was 'temporarily out of literature', but they hoped that I would join anyway. I did, for no particularly good reason, and a couple of months later found myself press-ganged into standing as a candidate in the 1977 GLC election. The party only had about 400 members in those days, a mere dozen of whom seemed to live in London, so candidates were obviously few and far between. So 'green' was I that I didn't even know that as an ILEA teacher, I was not in fact permitted to stand as a candidate for the GLC! It didn't seem to make much difference: 298 votes were clearly not enough to bring the full weight of the law crashing down upon me. Since then I've stood as a candidate in two local elections (legally), two general elections and two European elections, and been involved in more campaigns than I care to remember. As George Eliot wrote in her novel, *Felix Holt*, 'so our lives glide on, the river ends we don't know where, the sea begins, and then there is no more jumping ashore.'

Though the actual process of becoming involved may have been somewhat quirky, the impact on my life was far from such. In those days, apart from the fact that I loved teaching, I didn't really have a clue what was going on. Politically, I was almost professionally confused, agreeing with one party here, another there, and with none most of the time. I'd been to listen to the likes of Tony Benn and Denis Healey, but I'd also attended a couple of meetings of the Tory Bow Group, and I tended to vote Liberal. My thoughts about religion, philosophy and society in general were equally incoherent, and my lifestyle at that time may reasonably be described as aimless.

What has changed above all since then is that there is some consistency in all the things I do and believe. Ecology seems to have a way of either making complete sense or making very little sense at all. For me it was the former, affecting not only my politics but my whole way of life. I would hardly claim to be free of contradictions (and still find that it's the paradoxes of politics that often give me most pleasure) but green politics has given me an integrity, a wholeness, that was completely lacking before.

One last personal note. Many of the things I have learned and come to understand over the last ten years have been the result of the books and the magazines I've read. It has been this reading that really inspired me, made sense of things for me, and encourages me now to write this book. I lay no particular claim to outstanding originality, and a lot of what you'll find in this book has been dealt with by other authors far better qualified than myself. To them I owe more than the customary formal acknowledgement, for they have been my teachers and intellectual companions as much as anything else. I have quoted from them at regular intervals, and in other places have provided my own synthesis of what they have said before. At the end of the book you will find a very personalized bibliography in which I have tried to make clear the extent of the debt I owe them.

This is not just an apology for any inadvertent plagiarism. The more I find out about ecology, the more I realize that there is nothing new in what we are saying. Green politics today is the rediscovery of old wisdom made relevant in a very different age.

Part One

Seeing Green

I

Green Politics Today

It may be rather imprecise, but I shall be using the terms 'green politics' and the 'politics of ecology', or 'greens' and 'ecologists', more or less interchangeably in this book. I don't think it much matters which one uses, though personally I was delighted when the Ecology Party eventually changed its name to the Green Party in 1986. 'Ecology' is still a rather daunting word, perhaps too scientific, too specialized to convey the full scope of the green perspective. But then again, 'green' has its disadvantages too.

The word 'ecology' was first used by a German biologist, Ernest Haeckel, in 1870, but it wasn't until the 1930s that ecology assumed full professional status – the last of the sciences to do so. Professional ecologists study animal and plant systems in relation to their environment, with particular emphasis on the inter-relations and interdependence between different life forms. It is wholly appropriate therefore that the politics of ecology should concern itself with the interactions between members of one particular species (namely, us) and the impact that we have on our environment. There are still a few professional ecologists of a purist persuasion who resent such an interpretation – but you do *not* have to be the bearer of a professional qualification to be a good ecologist these days. Nor do you have to know the derivation of ecology (from the Greek words *oikos*, meaning 'house', and *logos*, meaning 'understanding' – which makes ecology the means by which we understand our planet), to qualify as a green!

The Environment Movement

The largest and the most easily recognizable part of the 'green movement' still consists of the various environmental groups with which this country is so prolifically endowed. In the past decade, membership of such groups is said to have risen from 600,000 to more than 3 million, a dramatic and most encouraging sign of public awareness. But one would be quite wrong to assume that *all* environmentalists are green, for there are almost as many shades of opinion within the environment movement as there are within politics itself.

I find it useful to think of three varieties of environmentalist, two of which sociologist Stephen Cotgrove describes in some detail in his book, *Catastrophe or Cornucopia*. The first are the conservationists, or traditionalists, who in many ways are the heirs to nineteenth-century liberal revolt against economic individualism and the utilitarian, materialistic values of that time. Conservationists today often tend to be rather similar, laying great emphasis on the restoration of order and traditional authority, albeit in small-scale, self-reliant communities. They would tend to see coercion through the law as an essential part of this process, being somewhat pessimistic about the possibility of necessary changes taking place of their own accord. They are not opposed to industrialism as such, and do not really want to change society – they just want to conserve the best bits of it that might otherwise go under. Such environmentalists tend not to be green.

At the other end of the scale are the radical, libertarian environmentalists, heirs to the very different tradition characterized by the anarchist ideals of Kropotkin, Thoreau and Godwin. They too are keen on small-scale, self-sufficient communities, but as a means of escaping from the bureaucratic, hierarchical pressures of contemporary industrialism. Personal autonomy is high on their list of priorities, and theirs is a decidedly optimistic assessment of human nature. They are opposed to the present industrial system, totally reject the idea that the present crisis can be solved by one technological fix after another, and are keen to adopt new lifestyles. For them, messing around on the margins of social change is simply not enough, for they seek a fundamental change in values. Such environmentalists usually are green.

Somewhere in between, and certainly in a majority, are the reformists, who are heirs to no particular tradition, but are deeply concerned about a whole range of environmental issues. They are basically centrists who, by and large, support the dominant social paradigm, and are therefore not opposed to industrialism as such and are rather nervous of any mention of 'fundamental change'. They tend to have conventional middle-class concerns and interests, and are mainly to be found amongst the ranks of the Labour, Liberal and Social Democratic parties – and right now would probably resent being called green, however fast they may be moving in that direction.

It seems quite clear that whereas a concern for the environment is an essential part of being green, it is, as we shall see, by no means the same thing as being green. That's why the Green Party does not call itself the 'party of the environment', though many have tried to fob it off with so convenient a pigeonhole. Frankly, the enormous differences that exist between many environmental groups render it utterly implausible that any coherent political programme could be put together on the basis of their contrasting attitudes. Moreover, it would be quite impossible to reach a broader constituency purely on the basis of a concern for the environment. There has therefore of late been a growing tendency to draw a much clearer distinction between environmentalism and ecology. Murray Bookchin, the American eco-anarchist, is famed for his diatribes against environmentalists, whom he sees as being guilty of a 'managerial approach to the natural world' that, far from challenging the existing authorities, merely helps them to do their job better! By trying to make an unworkable system work, they are merely exacerbating the problems; by refusing to distinguish between symptoms and causes, they prove themselves to be mere dilettantes wearing their hearts on their mottled green sleeves.

A Green Movement?

Apart from environmentalists, what then are the other constituent parts of this so-called 'green movement'? It's difficult to be too precise, but essentially there are three main types of green activist. There are those whose lifestyle or whose work may loosely be defined as green: smallholders, organic farmers, bicyclists, vegetarians, ardent recyclers, the 'small is beautiful' brigade, those into alternative medicine or appropriate technology, and many of those who work in co-operatives and the 'alternative economy'.

Then there are those in other pressure groups and campaigns whose whole approach to their activities has become increasingly green, particularly in the women's movement and the peace movement, but also among animal rights' campaigners, Third World groups, and those who oppose nuclear power, the arms trade and multinational companies.

And, last, there are the politically oriented greens. In this country, such people find themselves either within the Green Party or striving to do their bit within the established parties, as members of the Liberal Ecology Group (LEG), the Socialist Environmental and Resources Association (SERA) or even the Conservative Ecology Group (CEG), though its tiny membership is said to have become terminally confused in present circumstances. Since the Social Democratic Party (SDP) continues to see itself as all things to all people, it has not yet got round to setting up an official group, though the SDP Greens are unofficially beginning to get the message through to the right places.

Some have compared this surge of green awareness to the eruption of religious sects in the seventeenth century – the Shakers, Quakers, Diggers, Ranters, Pilgrims, Fifth Monarchists and Levellers. Their fiercely independent spirit of egalitarian politics, their love of the Earth, their decentralist tradition, and their passionate spiritual commitment, certainly number them among a long line of antecedents for contemporary greens. By contrast, Aurelio Peccei, founder of the Club of Rome, described such a combination of groups and individuals as forming 'a kind of popular army, with a function comparable to that of the antibodies generated to restore normal conditions in a diseased biological organism'.[1] Others have more earthily compared it with the way in which a vast and diverse profusion of plant life suddenly materializes on the top of abandoned dung-heaps: weeds, flowers, creepers, brambles, herbs – each different, yet each a part of an intricate, interdependent pattern of fertility and green growth. I leave it to you to work out what the abandoned dung-heap represents.

Now whether or not all this teeming greenery constitutes as yet a fully paid-up 'movement' is anybody's guess. Mine is that it *doesn't*, for the simple reason that for many of the thousands of people involved in green activities, the separate concerns they have still matter more than their sharing of a common green perspective. The whole protracted and often bitter debate about whether one should work for ecological goals through joining the Green Party, or working within an established party or sticking with single issue pressure groups, or simply doing one's own green thing, seems to me to be greatly exaggerated. For a start, different approaches suit different people and meet different challenges. On a personal level, it has been fascinating for me to work both with the Green Party and, more recently, with Friends of the Earth, a non-party political pressure group. Whilst remaining a committed individual member of the Green Party, I have come to appreciate that there are things FoE can do which the Green Party can't – and vice versa.

It's also true that different people are at different stages of awareness about the green perspective. Concern about individual issues often develops into a readiness to take on the totality of green politics. There comes a time for many people when a clarity of commitment is called for; though I'm filled with admiration for the efforts of SERA and LEG to cajole and even shame their respective parties into a greener outlook, I can't help but wonder what happens when loyalties become as divided as they must be in such organizations. To have been a 'green Liberal' at the last election and to have had to listen to David Steel yakking on in favour of Cruise missiles, renewed economic growth and all the rest of the paraphernalia resulting from the Liberals' uncomfortable alliance with the SDP, must have been a nightmare. I can sympathize, but only up to a point.

'Each to their own' may be a cliché, but to greens it really matters. The last thing we need is some kind of macho exercise to demonstrate who is the greenest of us all. All green activists suffer varying degrees of frustration, and yet we all contribute various skills, resources and experience with which to work away at the defences of our industrial society, some by tunnelling under the

actual walls to undermine them, some from outside with a
battering ram, some from within by making love to the soldiers
on the battlements. And some are already at work building a new
society within the ruins of the old.

The Green Party

There remains one major area of controversy, and that concerns
the different attitudes that green activists have as regards 'power'.
To some, the very mention of the word is unacceptable. Michael
Allaby, the well-known environmentalist, once wrote: 'Ecology
activists are not concerned with power: they have no wish to take
political or economic power from one section of society and give
it to another.'[2] I must tell Michael Allaby that as far as the
Green Party is concerned, that's just humbug. There are many
people in positions of power who hang on to that power only
through the ruthless exploitation of both people and planet. They
have no intention of relinquishing it – and from such people this
power must be taken. There are many ways in which this might
be achieved, and greens are involved in just about every one of
them, so long as they involve neither dishonesty nor violence nor
a betrayal of their ideals. It is true that greens show little interest in
the notion of retaining power as something to exercise over
others; their interest lies in the belief that *everyone* should be
empowered to determine the course of his or her own life within
the constraints of a finite planet. But those green sages who just sit
around waiting for some miracle to bring this about are going to
be sitting there a long time.

Likewise, eco-anarchists can carp on to their hearts' content
about the inherent evils of party politics and the compromises
involved in becoming a member of any large organization. It's
nice for some not having to live in the real world. All that matters
is that your party or organization or group should reflect your
ideals both in the conduct of its business and in what it stands for.
Theodore Roszak, perhaps the foremost advocate of the green
alternative in the USA, wrote of the green movement: 'No single
ideological formulation could possibly corral such a rich variety
of experience and feeling, and still preserve the independence and

authenticity of everyone's protest.'[3] I would agree with that, but add two comments: first, there's a great deal more to politics than protest and, secondly, there is no God-given requirement which says a political party has to base its activities on a fixed ideology. Having written the last two general election manifestos for the Green Party, I would be hard put even now to say what our ideology is. Our politics seems to be a fairly simple mixture of pragmatism and idealism, common sense and vision. If that's an ideology, it's of a rather different sort from those that dominate our lives today.

The role of the Green Party is obviously of considerable importance to the development of green ideas in this country. It's my contention that party political activity will always remain an essential part of that development; for better or worse, the Green Party is the only organization around prepared to take on that role to its fullest extent. But even as a political party, it has no illusions about the fact that its primary function is still an educative one, the spreading of green politics to as wide an electorate as possible. If it has one major complaint about some of its fellow activists in the 'movement', it would be their tendency to sit around analysing the green fluff in their navels instead of getting out there and telling the world there's a revolution on the way.

It just so happens that elections are an extremely practical way of getting one's message across. The Green Party's primary concern is at the local level, to start winning seats on local councils. Such is the power of our centralized media and political power structure that it cannot possibly achieve this without a simultaneous and equally strong commitment to politics at the national level – a commitment which it is now in a much better position to carry through. It's often bad for morale that the party continues to get a very low percentage of votes, but most of us have now realized that with our present electoral system, this is unlikely to change. It affects the Party's credibility in the eyes of some people, but does not seriously impair its ability to carry out its self-appointed role.

The second aspect of which is to operate as the conscience of green politics. (I would like to be able to interpret its role here as the political arm of the green movement, but it's difficult to be an

arm of something that still lacks a body!) In this respect, it is
able not only to put pressure directly on other parties, which has
proved effective in several instances, but indirectly to be of
assistance to greens in those other parties who are then able to
point out the awful grey deficiencies of their colleagues. This
party political function needs to be taken an important step
further: at a time when a green bandwagon is definitely beginning to trundle, and all and sundry are finding it suddenly very
opportune to book a seat on it, it is important that there is one
organization whose commitment to green ideals is not a matter of
fashion, of political expediency or of partial understanding. That
may sound very pious, but there is an integrity about green
politics that fits ill with the machinations of contemporary
politics.

There has to be a balance between defending that integrity and
being open and flexible enough to encourage people to join the
greens rather than frightening them off. Such a balance requires
the drawing of a line, and we all draw it differently according to
our perception of contemporary politics. For me personally, *the
minimum criteria for being green* would run roughly as follows:

a reverence for the Earth and for all its creatures;
a willingness to share the world's wealth among *all* its peoples;
prosperity to be achieved through sustainable alternatives to
the rat race of economic growth;
lasting security to be achieved through non-nuclear defence
strategies and considerably reduced arms spending;
a rejection of materialism and the destructive values of
industrialism;
a recognition of the rights of future generations in our use of all
resources;
an emphasis on socially useful, personally rewarding work,
enhanced by human-scale technology;
protection of the environment as a precondition of a healthy
society;
an emphasis on personal growth and spiritual development;
respect for the gentler side of human nature;
open, participatory democracy at every level of society;

recognition of the crucial importance of significant reductions in population levels;

harmony between people of every race, colour and creed;

a non-nuclear, low-energy strategy, based on conservation, greater efficiency and renewable sources;

an emphasis on self-reliance and decentralized communities.

Some will find that too exclusive, and some will say that it doesn't go far enough. Many would no doubt add that it's a somewhat idealistic set of criteria, to which I would have to plead guilty. Despite all their experience, and all the evidence that confronts them, most members of the Green Party remain idealists. This is the only way I can account for the ultimate paradox about its existence: for despite being a national political party, intent, as I have strongly asserted, upon changing the balance of power in this country, very few of us realistically see its future in terms of growing numbers of Green Party MPs clustering on those green leather benches at Westminster. At the back of our minds, it's obvious to us that within the next generation, *all* politicians and *all* parties will *have* to become more or less ecological in their outlook. If they don't, then it's doubtful whether the trappings of democracy will be there for anyone to enjoy anyway. Given those assumptions, and the prospect that much of our politics would then take place at the local level without all the present clutter of party political labels, the Green Party *in its present form* should only have a limited lifespan. Not only is this a highly ecological recognition of the eternal cycle of life and death, but it must also make it one of the few political parties in history whose success will be measured by the speed with which it puts itself out of business!

Europe and the Greens

There is little doubt that the Green Party has benefited considerably from the success of green politics in Europe; an element of 'transferred credibility' has done wonders for us. It is also something of a comfort, when the going gets rough, to know that we are not on our own, but merely one part of a much broader, international movement. There are now green parties in West Germany, Belgium, France, Ireland, Sweden, Austria, Luxemburg, Switzerland and Holland, as well as in the UK. Green groups in Finland, Spain, Portugal and Greece are planning to become parties in the near future. In 1979 a co-ordination group of European green parties was set up to exchange information, develop a common manifesto for the 1984 European election, and plan joint actions.

There is no doubt that the influence of the greens is spreading rapidly in many European countries, but especially in Belgium and West Germany. In November 1981, the two Belgian parties (Ecolo, the Walloon or French-speaking party, and their Flemish counterparts, Agalev) had five senators elected to the Upper Chamber and four representatives elected to the Lower Chamber of the Belgian Parliament. This was the first time a specifically green party had had MPs elected to any national parliament. In 1982, they won 120 seats on local councils, and now hold the balance of power on three of them.

By far the best-known of the European green parties is die Grünen – indeed, there's far more coverage of the German greens in the UK press than there is of the UK greens, which is not saying much, since there's more coverage of us in the *German* press than there is in the UK press! – and it was their success in March 1983, winning twenty-seven seats in the Bundestag with 5.6 per cent of the vote, which firmly established green politics as a serious alternative, and inspired green growth everywhere.

Die Grünen were not formally constituted as a political party until January 1980, and the way this came about is an interesting example of the very close links between green parties and broader green movements. It was the issue of nuclear power which served as a catalyst for the emergence of die Grünen as a political party. In the mid-1970s, most environmentalists in Germany were confident that nuclear power could be stopped by non-violent direct action, including the occupation of new sites. However, their experiences at places like Whyl and Brokdorf led them to realize that such actions, divorced from the exercise of real political power, were quite inadequate. Other approaches were tried, including application to the courts and efforts to win round the Social Democrats (SPD) to an anti-nuclear position. Neither was particularly successful – which will hardly come as a surprise to greens in this country.

At that stage most of the *Burgerinitiativen* (a coalition of environmental and citizen action groups) were still opposed to setting up a separate green party, but from 1977 onwards green candidates began to stand in local elections. The continuing failure of all other strategies led to renewed calls for a green party, so that extra-parliamentary efforts could be backed by radical parliamentary involvement. The environment movement became increasingly politicized (something that has never happened in this country) and began to realize that getting rid of nuclear power would only be possible as part of a much broader political and social transformation. A joint platform between the different groups was rudimentarily cobbled together for the 1979 European elections, and in the next year the party was formally set up.

Over the next three years they consolidated their ideas and established a broad membership. They became far more involved in the peace movement, were able to exploit the dissension in the SPD, and rapidly became known as the 'Peace Party'. Their uncompromising stance on Pershing II and Cruise missiles made theirs the only voice of authentic opposition in West German politics. As a result, they had considerable success in the local elections, and won several seats in various regional parliaments. It was here that they established their credibility and prepared the groundwork for their election to the Bundestag in 1983.

A very different story from that of the British Green Party! It highlights many of the reasons why die Grünen have enjoyed such success: a well-established base at the 'grass roots'; a politicized environment movement; a passionately fought campaign against nuclear power; total commitment to a peace movement not dominated by the left; a federal parliamentary system; and, above all, an electoral system (with representation allocated proportionately once a 5 per cent threshold has been crossed) that has allowed a green party to claim its rightful place alongside the established parties.

Since their election to the Bundestag, things have not been easy for them. The fact is that right from the start, die Grünen have been a rather uneasy coalition of many different political view-

points and ill-fitting ideological hang-ups. Very damaging internal divisions take up a great deal of their time. Much of their
support comes from the political left, including an influential and
powerful group of ex-communists. As a result there is now a very
a serious split between the 'reds' and the 'greens', and the present
Executive of die Grünen has done little to endear itself to other
European green parties by forming ad hoc 'oppositional alliances'
with any old radical, progressive political grouping – including
official Communist Parties. There is an equally important divide
between what might be described as the 'fundamentalists' and the
'realists'. The fundamentalists want to move as fast as possible
from a capitalist to an ecological society, and see their main task as
that of representing extra-parliamentary movements within the
Bundestag. The realists are more concerned about gradual
change, and are prepared to contemplate major compromises as a
means of widening their electoral base. There is much discussion
of the possibility of an 'understanding' with the SPD, and some of
the regional parties have already taken steps in that direction.

Yet it strikes me that Petra Kelly, whose integrity, courage and
impassioned commitment to *genuine* green politics have justly
won her international renown, is right when she asserts that die
Grünen can only gain by refusing to compromise at this stage. If
there is something unique about green politics, why clutter it up
with short-term expedient trade-offs with either the left or the
centre? The left would like nothing more than to co-opt the
vitality of the greens to promote its own faded dreams. And the
SPD would like nothing more than to co-opt the growing
support for die Grünen as a way of winning back all its lost voters
without genuinely having to confront the challenge contained
within green politics. This is something greens in this country
would do well to think about, as one watches the wheelers and
dealers in both the Liberal and the Labour Party trying to lay
claim to the green vote. Despite any number of temporary
difficulties, green politics has something totally different to say
and a totally different way of saying it, both here and in
Germany. We are *not* open to co-option, for the tide is running
our way.

2

Opposing World Views

Across the Divide

The more successful the greens become, the harder their task will get, for so much the greater will the backlash be. So powerful is the opposition to any real change, so little scope is there for achieving progress through consensus, that greens have inevitably been forced into the politics of conflict, shifting from reform to radical challenge. There are enormous differences between our goals and values and those of society at large. That there are two totally different world views at work in contemporary society will become clearer as we move through the analysis of industrialism in Part Two to the presentation of a green alternative in Part Three. What is rational and reasonable from one perspective is often deeply irrational from the other. If one's goal is the maximization of economic output as a means of keeping people happy, then such things as nuclear power, built-in obsolescence and involvement in the arms trade are all 'rational'. If one's goal is a more convivial, sustainable society, then such things seem increasingly barmy.

From our side of the divide it's clear that all industrial nations are pursuing an unsustainable path. Every time we opt for the 'conventional' solution, we merely create new problems, new threats. Every time we count on some new technological miracle, we merely put off the day of reckoning. Sheer common sense suggests alternative remedies, yet vested economic interests and traditional political responses ensure that the necessary steps are never taken. The old system endures, dominated by competition between various groups struggling for power so as to be able to promote the interests of a particular class, clique or ideology. Such power is retained by the careful use of the legal stick and the financial carrot. The horizon is the next election.

But one can't help feeling sorry for politicians – of all persuasions. They are just ordinary people who must today be totally bewildered, not quite knowing why their gods have forsaken them, why policies that once seemed so sensible simply don't work any more. They can see the evidence all around them of a growing crisis of confidence in our industrial civilization. They must also realize that their politics is threatened at its very foundations – namely, in its ability to satisfy people's material expectations. They are slowly and agonizingly being hoist with their own petard: having used economic criteria as the sole benchmark for establishing success, value or achievement, they are now compelled to search desperately, without any real hope of success, for new economic miracles.

The irony is that many politicians agree with much of what we say, but cannot admit this openly or act accordingly. There is a conflict between private principle and public practice. The majority, of course, still *don't* agree, and even seek to belittle this opposing world view. The trick here is to question our sanity, or to make out that we're all a load of emotional misfits. Paul Johnson refers to the ecology movement as 'simply irrational'. In his 1978 Dimbleby Lecture, Lord Rothschild dismissed us as 'eco-maniacs' and 'eco-nuts'. Whenever at a loss for a cheap jibe, something along the lines of 'woolly-hatted, woolly-minded lentil stirrers' is readily available, though it's often couched in rather more formal language.

E. F. Schumacher used to claim that he never minded this sort of abuse, and in particular enjoyed being called a 'crank', since a crank is a small, simple, inexpensive and efficient tool – 'and it makes revolutions'. None the less, it is worrying when someone like Professor Ralf Dahrendorf, who is usually so perceptive about social and economic trends, can write: 'The Greens are essentially about values, an imprecise, emotional protest against the allegedly overbearing rationality of the social democratic world.'[1] Professor Dahrendorf's understanding of green politics is still, by his own acknowledgement, very patchy; but to fall into the trap of dismissing something as 'emotional' merely because its assessment of rationality does not accord with one's own is a sad lapse from grace.

Reason, Sweet and Sour

That the green perspective *does* have a different view of reason or rationality is hardly surprising, for its critique of industrial society is radical in a remarkably literal way. By refusing to abstract our human concerns from the web of life that is our biosphere, it seeks to examine the very roots of human existence. Simply 'getting the facts right' is a much more complex business than today's rationalists would have you believe, and no judgement can ever be 'value-free'.

In his quite excellent pamphlet for the Green Alliance, *Economics Today: What Do We Need?*, eco-philosopher Henryk Skolimowski reminds us of the memorable scene in Bertolt Brecht's play *The Life of Galileo*, where Galileo pleads with the courtiers and the scholars just to look into the telescope and see for themselves the proof that the world goes round the sun rather than the sun going round the world. But they wouldn't, for they were simply incapable of coping with any new facts that might overthrow the existing order of things. After all, they saw themselves as 'defenders of the faith'. Kids today are always taught to admire Galileo and his courageous stand against the reactionaries and religious bigots who put him on trial, without so much as a passing awareness that it is now the world view of Galileo and others like him that has become the dominant orthodoxy, and that is now suppressing a different vision and a different interpretation of human destiny.

This dominant world view, the consequences of which we shall consider in detail in the next part, has come to the end of its useful life, not least because its notion of rationality is so woefully lacking. We like to think that ours is the supremely rational civilization, but does that claim really stand up to any kind of examination? Instead of looking into space through Galileo's telescope, let us look down at ourselves. Imagine, if you will, the proverbial little green person from Mars taking stock of Planet

Earth, and the UK in particular. Having anticipated a model of
rationality, might 'our Martian' not be surprised to find:

> that it's apparently possible to keep the peace only by threaten-
> ing the total annihilation of the planet;
> that it's possible to achieve 'progress' and further growth only
> by the wilful destruction of our life-support systems;
> that so civilized a nation bats not an eyelid as it inflicts terrible
> suffering on its fellow creatures;
> that we obsessively promote the most expensive and most
> dangerous energy source to the exclusion of all others;
> that millions remain unemployed when there's so much
> important work crying out to be done;
> that millions more carry out soulless, mind-destroying jobs
> that make nothing of their resources and creativity;
> that we ravage our best farming land to grow food surpluses
> that are then thrown away or sold off cheap to the 'enemy';
> that we consider the best use of the proceeds of North Sea oil is
> to keep people on the dole;
> that in one part of the world millions die of starvation, while
> people here die of over-indulgence;
> that we spend as much on useless weapons of war as we do on
> either education or health;
> that our 'planners' have allowed rural communities to waste
> away, while making inner-city areas uninhabitable;
> that we pollute the planet in the very process of trying to get
> rich enough to do something about pollution?

Now that's just the first 'dirty dozen' out of our little green
Martian's notebook, and the list would go on and on and on. It's
hardly a prima facie case for a rational, civilized society. So bear
with me if I go on a bit about just who is rational and who is
irrational in this crazy world of ours. Ecologists get very emotional about rationality! We've had our fill of the Rothschilds and
the Paul Johnsons, who disparage what they can't comprehend
and mock what they can't live up to. Though we would never be
so foolhardy as to assume that reason alone is sufficient to build a
caring, civilized society, the politics of ecology is none the less
profoundly rational.

I would, however, be the first to agree that the way in which we express that reason may at first be a little unfamiliar. I was lucky enough to be invited to the opening of the West German Bundestag in 1983, as part of a delegation of European greens. After a tremendous fuss about seating arrangements in the actual chamber, the twenty-seven Green MPs had been placed, two by two, in a long thin row between the two major parties. They had decided that they would wear their ordinary clothes in the Bundestag, and that for the opening they would each take in some plant or small tree to show their commitment to the politics of life. When I first heard that, my heart sank, presuming in my 'rational' way that they would look silly; jokes about 'flower-pot people' were surely the last thing we wanted to encourage. And yet, when I saw them in that drab, lifeless chamber, surrounded by hundreds of men and one or two women all identically dressed in sombre black, they made such a splash of vibrant, living colour as to make my earlier scruples appear mean and trivial. There was *nothing* irrational in what they did, for as greens our primary concern is to remind people of the inseparable links between ourselves and the planet on which we depend.

What Price Progress?

These are difficult times for people; the old is wearing thin and many can't quite see where the new is coming from. There is much to explain why people should hang on to what once seemed to work rather than take a leap in the dark in the direction of a very different future. For the Age of Industrialism has indeed brought enormous benefits to millions of people, in terms of both material improvements and democratic rights. It serves little purpose to deny that enormous progress has been made over the last two hundred years, and there is certainly no question of the politics of ecology harking back to some pre-industrial Golden Age. It wasn't golden; it was often mean, miserable and moronic. Nor is it particularly worthy to deny the humanitarian intentions of those whose visions of material plenty and the elimination of poverty and oppression now begin to look so threadbare.

The trouble is that we have simply taken their particular

interpretation of progress for granted, without realizing that the very forces which have enabled us to make such improvements could, as Roszak puts it, 'overshoot their promise' and bring about a 'new dark age'. We have not been sufficiently sensitive in realizing that many of the obvious advantages of technological change also carry a certain price. Motor cars, drugs, pesticides, TV, the very process of mass production: the contribution of each of these to the pattern of progress is undeniable, but we have been remiss in overlooking the damage they have done. We have become befuddled by change itself, with a powerful assumption that without rapid change, there can be no progress.

And that is where ecologists part company with supporters of the present order. The burden of proof should not be on us. We should not have to produce evidence of the damaging consequences that may result from technological progress – it should be the other way round, not only because we can now weigh more clearly the benefits *and* the costs of industrialism, but also because from now on, at a time of growing ecological tension, there must be a presumption in favour of the planet and against our aimless, uninformed impact on it. That is why I have called Part Two 'The Balance Sheet', for up until now our accounting procedures have been woefully inadequate. The benefits are writ large for all to see in our self-congratulatory society, but the costs are all too often ignored, concealed or written off in one way or another. To use the jargon, we have been involved in a double 'trade-off' that has affected every corner of the planet. First, we have assumed that important social goals like full employment and the creation of wealth are not compatible with the protection of the environment. If we want these things, so the argument goes, we must pay the price for them; so the environment and the maintenance of our biological support systems have been traded off against material progress. Secondly, we have traded off the future against the present; short-sighted cost/benefit analysis, with its emphasis on present value and instant reward, has discounted the future and the long-term effects of our current activities.

To us, these trade-offs appear staggeringly foolish – and yet they remain at the heart of industrialism. The idea that people's welfare can be promoted by systematically devastating the environment on which ultimately we all depend, and by ruthlessly sacrificing the future regardless of the interests of future generations, can only be described as lunatic. How is it that such profound irrationality has taken such hold of our actions and words as to undermine our security, fundamentally impair our quality of life and now threaten our very survival?

Part Two

The Balance Sheet

3
For Earthly Reason

It's no doubt rather traditional for an ecologist to start off any analysis with a review of the state of the planet; yet not to do so would indeed be peculiar, since the single largest impetus in the formation of a radical green front has been the failure of the established political parties to do anything about the care and maintenance of the planet. The state of the planet provides the context within which *all* politicians operate. Yet the vast majority of them remain oblivious of that context, or choose to ignore it. For those involved in green politics it is this context that gives shape to everything we say and do. It provides the framework within which we develop our ideas, dream our utopias, amend our lifestyles. It provides the earthly reason for all our labours.

Since the early 1970s there has been a steady stream of international reports and 'global overviews' of the state of the planet. They have done much to raise public awareness and to focus attention on this simple message: 'If present trends continue, the world in 2000 will be more crowded, more polluted, less stable ecologically and more vulnerable to disruption. Despite greater material output, the world's people will be poorer in many ways than they are today.'[1] But their impact on world leaders has been all but imperceptible. Even *Global 2000*, the authoritative and prestigious report commissioned specially by President Carter, is now just another global dust-gatherer. For most of them have one major fault: they just hang there, suspended in a gravity-less atmosphere, untouched by the realities of contemporary politics. The *World Conservation Strategy* provides an interesting exception in this respect, but it is so deeply flawed by the laboured efforts it has to make to bridge the divide between two conflicting world views that it ends up satisfying neither.

The 'limits to growth' debate still simmers away, with arguments about when such and such a resource will run out, or when the finite physical limits of the planet will be reached. It's now obvious that such arguments are largely irrelevant in themselves; the present industrial system is not immediately under any threat because of things running out, and there's no doubt that global exhaustion of most resources is still some way off. The doom-laden warnings of the 1970s have not proved accurate, for necessity is indeed the mother of invention. It is rather the *combination* of so many different factors that provides the element of urgency. Each 'crisis' or threat examined on its own may well be capable of solution through fairly conventional approaches, but when all these threats combine together, their effect is to shatter that whole conventional superstructure.

Pop Goes the Planet

In framing an earthly context for contemporary politics, it makes sense to start with population – if only because so many politicians go to such lengths to avoid discussing it. There seems to be something about the sensibilities of all good liberals that makes it extremely uncomfortable for them to cope with population matters. It's obviously a problem, so obviously something's got to be done; but exactly what, by whom and in what way are emotive and controversial areas. So it remains a taboo subject.

The bare statistics have become familiar: a world population two thousand years ago of about 250 million, the first billion reached by 1830, the second a hundred years later in 1925, the third thirty-seven years later in 1962, and the fourth just thirteen years later in 1975. World population is now about $4\frac{1}{2}$ billion, and it will be 6 billion by the end of the century. Population growth has been a brief, abrupt phenomenon, a veritable explosion. Professional ecologists draw a comparison with other species which experience explosive phases of growth when there is an abundance of resouces and a relative absence of predators and disease. Such a growth phase is usually followed by a period of contraction, till the numbers are reduced to a level which can be sustained by the environment. The question is, are we just another species?

At the most basic level, the answer is 'yes'. No species can repeal the self-evident law that indefinite increases in population are simply not sustainable on the basis of finite resources. And there's already a fearsome momentum built into population growth on account of the relative youthfulness of some countries; 45 per cent of all Africans are under fifteen years old, as are 40 per cent of Latin Americans and 37 per cent of Asians. The United Nations (UN) and the World Bank talk of world population stabilizing at 10 billion some time in the next century. They also estimate that in twenty years' time, the world's population will be divided equally between living in cities and living in rural areas, on account of the huge migration of the rural poor into the cities. It's hardly possible to contemplate such eventualities. As the 'revolution of rising expectations' combines with population growth, it will impose massive pressure on the Earth's biological support systems.

The great myth, of course, is that overpopulation is exclusively the fault and the concern of the Third World. On a planet as interdependent as ours, this would be questionable at the best of times; given the current excesses of affluence and profligate use of resources that characterize the developed world's standard of living, it becomes a transparent lie. In the UK we eat on average three times as much and consume forty times the amount of fossil fuels and industrial products as the average citizen of the Third World. Moreover, we ourselves live on an extremely overcrowded island; the density of population in England is four times that of China! We remain dependent on imports for 50 per cent of our food and animal feed and, apart from energy supplies, have relatively little by way of raw materials; our continuing balance of payments problem is just one of the penalties we pay because our population is far in excess of available resources. It is generally agreed that the optimum size for this country would be around 30 million. With our current population at 56 million, we too need to ensure that our birth rate keeps falling – and falling much more dramatically.

The other great myth is that the problem of overpopulation is greatly exaggerated and that it's inappropriate to push the Third World too hard in this respect. Those who hold to this myth are a strange bunch, made up of extreme left, extreme right and the rump of reactionary Catholicism. Those on the left correctly point out that the reason why there's world hunger is not because there's any real shortage of food, but because it's unfairly distributed. Up until now food supplies have indeed increased faster than population, though not without exceptions: Africa's per capita food production has fallen 11 per cent since 1970. But such an analysis is dangerously limited, inasmuch as it implies that a more equitable distribution would alone solve the problems of overpopulation. In her otherwise incisive critique of the Brandt Report, *The Creation of World Poverty*, Teresa Hayter writes: 'It is not clear how much rapid increases in population do, in fact, add to the difficulties in providing reasonable standards of living.'[2] The irony is that so unfortunate a lack of clarity leads to a position not so dissimilar to that of extreme right-wing mavericks like Julian Simon, in whose terrifyingly irresponsible book, *The Ultimate Resource*, we hear a passionate plea for more rather than fewer people.

Of a different order, because of its moral and religious justification, is the official position of the Roman Catholic Church. For all that the present Pope is a wonderfully charismatic leader of people, he's not doing God's world any favours by his attitude towards birth control. A belief in the 'sanctity of life' is often raised in this context, but it strikes me that this has become such an absolutist position as to allow of little humanity and even less ecological wisdom. There is a peculiar callousness in those who can so resolutely ignore the appalling suffering caused by unwanted pregnancies, unwanted births and overpopulated environments. By opposing access to family planning facilities, and by denying the right to legal abortion, the 'sanctity of life' lobby must take their fair share of the responsibility for the danger and pain which millions of women have to endure, and for the degradation and suffering experienced by millions of children.

Even if we're optimistic, and err on the side of caution in

projecting population levels, it's clear that the path we are on is inconsistent with the evolution of a sustainable society. We must therefore reject the notion that the world's population will automatically double in a generation. We must realize that the current proportion of aid spent on family planning projects is quite inadequate and, contrary to another widespread belief, completely fails to meet the demands of Third World countries themselves. As we shall see later, there are many things that can and should be done, but a combination of national commitment and international support is essential for all of them. As Eric McGraw says in his pamphlet *Proposals for a National Policy on Population*, 'This is an issue intolerant of government pressure – and yet endangered by government procrastination.'[3] The danger is incontrovertible: only as world population moved towards the 4 billion mark did it begin to outpace the production of basic commodities. If it increases at the same rate, the pressure on the Earth's biological resources will be literally unbearable. In some respects, it already is.

The Earth's Epidermis

The very idea of 'topsoil' has always fascinated me. There they are, these billions of micro-organisms, beavering away to build up the fertile layer of soil that makes it possible for us to live. I can just imagine David Bellamy waxing lyrical about so miraculous a process! And I would listen entranced, for it *is* a kind of miracle.

But this thin outer layer, rarely more than a foot deep, is under great pressure. An inch lost through erosion may take centuries to replace, and though erosion is a natural process, there's real trouble whenever the rate of erosion exceeds the rate of soil formation. Modern farming techniques are threatening to achieve just that. World food output has doubled since 1950, but this has only been possible through the abandoning of traditional farming methods (such as rotation and fallowing) and the input of massive amounts of cheap fertilizer. Monoculture, the planting of just one crop year after year, field after field, has replaced the more diverse planting patterns of earlier times.

Lester Brown, one of the most ardent and articulate defenders of the soil, provides some horrifying statistics. In the state of Iowa alone, 260 million tons of topsoil are lost every year from cropland. Land abuse has been so severe over the last few years that fully one fifth of the world's cropland is losing topsoil at a rate that is undermining its productivity. A 1980 report from the UN's Food and Agriculture Organization reported that between 13 and 17 million acres of cropland were being completely lost for agricultural production every year – at exactly the time when production will need to double again to feed the world's increased population.[4] Combine this with the process of desertification, through which it is calculated that 50 million acres are lost annually,[5] and the extent of the problem becomes all too apparent.

There are no easy answers. Consider the dilemma of a modern farmer. Erosion control measures are costly and, in today's economic jargon, not 'cost-effective'. If his profit margins are low, the farmer only has two choices when confronted by excessive erosion: to introduce controls, lose money and possibly face bankruptcy; or to carry on 'mining the soil' till fertility falls so low that the land has to be abandoned. Or consider the dilemma of planners and government officials. The cities spread and spread, and with them the need for increased building and road construction. In the USA, nearly a million acres of prime cropland is converted to non-farm uses every year; in this country we lose about 125,000 acres. The land is expendable; the people must go somewhere – or so the argument goes.

With most of the world's cropland already under the plough, the potential for expansion is severely limited. At the most, another 10 per cent may be added to the world total. That makes the present loss of soil all the more desperate. There may be substitutes for oil, but there are none for earth; Lester Brown puts it this way: 'Croplands are the foundation not only of agriculture, but of civilization itself... the loss of soil is in some way the most serious of the threats civilization faces. It can survive the exhaustion of oil reserves, but not the continuing wholesale loss of topsoil.'[6]

Disappearing Forests

Nor, indeed, the continuing wholesale loss of our forests. It so happens that I'm writing the first draft of this book surrounded by 60 acres of trees, most of which I planted myself when we used to live out here in New Zealand. (That was long before I joined the Green Party!) They are now fifteen years old, and it's a good place to be. Trees have always mattered to me, just as they matter to millions of people, yet it's easy to forget just how dependent we are on them, for commercial and aesthetic reasons, for biological and genetic reasons, and, as regards a large proportion of humanity, for cooking and home heating. Given such dependency, it's hard to exaggerate the irrationality of our current exploitation of the world's forests. The recent pattern of deforestation, coupled with totally inadequate replanting and soaring growth in demand for many timber products, is a recipe for guaranteed disaster.

A lot of attention has focused on the world's tropical forests, which in many countries have been exploited just as fast as has been commercially possible. The UN's Environment Programme's recent survey indicated that more than 18 million acres a year are being cut down. It sounds appalling, and it is – although it should be remembered that the Amazonian rain forest covers 700 million acres, three times the size of France! The scale of such losses may be less than some people had anticipated, but in the absence of any sound management or land-use plans, grave ecological damage is being done and future economic potential lost. South-East Asia has suffered even more than South America, and one can only regret the large amounts of international aid that were made available to finance replanting schemes in such a way as to promote private profit and encourage further foreign consumption.

It is hardly surprising that the distribution and use of forest products should so accurately reflect the global distribution of wealth. In each year the average US citizen consumes about as much wood in the form of paper as the average citizen in many Third World countries uses for firewood! Most of the consumption of what is called 'industrial' wood (a phrase that has a predictably ugly ring to it) takes place in the developed world, though most of the forests grow in the Third World. Eighty per cent of wood used in the Third World is burned for fuel, and it is here that the crisis has become most urgent. Millions live in areas where the amount of wood needed far outpaces new growth. This imposes a real burden, mostly on the women who have to collect the wood from further and further away. Prices have soared, and considerable economic hardship has resulted. The increase in new planting that will be necessary to meet this demand is enormous, and in the meantime many villagers have little choice but to continue to wreak great damage on their own environment. As Erik Eckholm says, 'Uncontrolled deforestation is a symptom of a society's inability to get a grip on other fundamental development problems: agricultural stagnation, grossly unequal land tenure, rising unemployment, rapid population growth and the incapacity to regulate private enterprise to protect the public interest.'[7]

Black Magic

There's not a lot more to be said about oil, really. There was once a lot of it, there's a lot less now, and in the year 2000 there will be a very great deal less. There's still some controversy about *exactly* how much is left, but most experts more or less agree on figures of 650 billion barrels of proven reserves and about 2,000 billion barrels of 'ultimately recoverable reserves' – give or take a few billion barrels. That sounds a lot, but what it means is that at the rate of consumption of the average American, namely thirty barrels per person per year, it would be all gone in about *fifteen years*. Fortunately, the rest of the world can't touch the Americans for sheer unrestrained excess in this respect. Annual global consumption is now about 19 billion barrels, down 15 per cent from the peak of 1979, and it's expected to decline by another 15 per cent by the turn of the century. Given the expected increase in population, average consumption will therefore be down from five barrels per person to two and a half barrels per person. One doesn't need to spell out just how drastically that's going to affect every single person on this planet. There may be temporary upturns, especially over the next couple of years, after so prolonged a recession, but the trend is clear and undeniable. 'Depletion psychology', coming to terms with the day that reserves will start running out, will become a feature not only of the Organization of Petroleum Exporting Countries (OPEC), but also of the life of the average consumer.

One consequence that is little discussed is the impact this will have on land use. In the past, cheap oil has subsidized agricultural production through mechanization and cheap fertilizers. But as the price of oil rises, so will the price of oil-based agricultural inputs and farming techniques. In the past, oil has substituted for certain natural materials, through the use of plastics, synthetics, etc. But as the price of oil rises, such substitutions may be reversed, thus putting more pressure on cropland. Energy requirements are already beginning to compete with food production in a much more direct way. Alcohol distilled from agricultural products is seen by many countries as the best alternative to imported oil. Cars compete with people in terms of land use, and although a country like Brazil, the front-runner in this development, says that the crops produced for this purpose will be in addition to rather than instead of crops for food, one can't help being a little cynical about it. After all, cars have more purchasing power than the average Brazilian. To meet its aim of being self-sufficient in liquid fuels by 1990, Brazil has plans to plant sugar cane on an area half the size of the total available for growing crops. It's hard to see how the poor won't lose out yet again.

Food for the Future

Agricultural productivity remains the key to what's going to happen in the future. Over the last ten years, increases in agricultural production have barely kept pace with population growth. With 70 million new mouths to feed every year, and very little new land to be found, greater yields have got to be gained from existing croplands. Fertilizers have been the principal source of such growth in the past, but as the amount of cropland per person shrinks, so the amount of fertilizer will have to be increased. We've already seen why that may well not be possible. More than a hundred countries today are dependent on North American grain. Many erstwhile exporters of food have now become importers; it never seems to happen the other way round.

At the moment a substantial proportion of the annual grain harvest is fed to livestock; when supplies of grain run short, the competition between people and livestock becomes intense. Again, the purchasing power lies with the more affluent, and that means the livestock get the grain so we can get our hamburgers. It's an exceedingly inefficient means of producing protein at the best of times, and exceedingly inequitable at the worst. The fact remains that the grain that now goes to satisfy the meat eaters of the developed world is the only theoretical reserve that exists in the event of any serious shortage of food elsewhere.

It's hard to keep track of all the different factors that are going to affect the price of food. As I said at the start, any one could be dealt with in conventional terms. But looking at the following list, the combination becomes devastating: an increase in world population; an increase in consumer expectations; an inevitable rise in the price of oil; an increase in the cost of fertilizer and a decline in its use; an increase in the costs of all energy-intensive inputs; a loss of cropland to urbanization; a loss of land to soil erosion and desertification; cropland going into production of grain for energy uses or to replace synthetics with natural materials. Is it any wonder that conventional economists have little to say, and have consciously or unconsciously settled for the fact that famine and malnutrition can only get worse?

Some anticipate that more food will be taken from the oceans, but here again a similar combination of problems would thwart their naive optimism. The world catch tripled between 1950 and 1970, but only as a result of the most ruthless overfishing. From 1974, it started levelling off, and it seems highly unlikely that the growth era will ever return. The story of the Peruvian anchovies, though often told, retains a grim fascination. It reads like a parable warning of the folly of the modern world. There were once huge concentrations of the humble anchovy to be found off the coast of Peru. Aware of the very considerable demand for fishmeal in the developed world, to feed its chickens and pigs, the Government built up the world's biggest fishery to exploit this marine bounty. Quite soon one-fifth of the entire global fish catch was hauled up out of those Peruvian waters, and lots of people got very rich as a result. But they also got greedy. Despite warnings from eminent ecologists, contemptuously dismissed as prophets of doom, for five years running the annual catch exceeded the maximum levels that had been agreed on. People got even richer. And they also got greedier. But in 1972 came retribution! No more anchovies! Empty nets! Fortunes lost! And from that day to this the anchovies have never returned in their former numbers.

It's a sad story, but one that has been repeated time after time the length and breadth of the planet. We not only destroy its riches through such heedless exploitation; we cut our own throats.

Gross National Pollution

Everyone's against pollution – in theory. It's like everyone's against waste – in theory. And it's true that there have been some significant improvements over the last decade that in certain areas have made a real impact. However, ecologists are not deluded into supposing that the hauling of one unfortunate salmon from the murky waters of the Thames, for the first time in years, marks the turning of the tide. For the basic problems of exactly how we cope with pollution still remain.

The *costs* of pollution control are easily quantifiable; there they stand in the books as a debit. The *benefits* are not as easily quantifiable, and society's ledger doesn't have a special column to show the advantages of a cleaner environment. Since the direct expenses fall on particular industries, they have a very strong political interest in opposing pollution control; since the benefits are spread much more widely throughout society, there is often no single group that can take up the cudgels with the same single-minded enthusiasm as industry does. A classic example of this has been the case of lead in petrol. Only after many years of campaigning did the Government eventually yield in 1983 to the combined onslaught of the Royal Commission on Environmental Pollution and the pressure group CLEAR. A decision was

taken to remove lead from petrol 'before 1990', but the prevarica-
tion goes on even now, and as yet nothing has been done to
implement the decision.

It is this disequilibrium of power which compels us to dismiss
the dangerous yet ever popular notion that nothing can be done
about pollution unless the economy is expanding. Fritz
Schumacher quotes a wonderful example of this wilful drawing
of vicious circles from a certain Professor Heller, former Chair-
man of the US President's Council of Economic Advisers: 'We
need expansion to fulfil our nation's aspirations. In a fully
employed, high-growth economy you have a better chance to
free public and private resources to fight the battle of land, air,
water and noise pollution than in a low-growth economy.'[8] Since
pollution is one of the direct results of a high-growth economy,
where, might one ask, is the circle to be broken? It's like trying to
give up beating your husband or wife by beating them twice as
hard. An increase in Gross National Product (GNP) *inevitably*
means an increase in Gross National Pollution.

There's also the problem of ecological time-lag, the fact that
some damage becomes apparent only when it's much too late to
do anything about it. This occurs either because we fail to
monitor things, or because we don't know how to monitor them,
or because we fail to take action even when we do – as in the
scandal of the asbestos industry. The whole fabric of our daily
lives is made up of industrial products or chemicals of one kind or
another; 55,000 chemicals are produced commercially, with 1,000
new ones coming on the market every year. Our thoughtless
acceptance of these products in the past means that only now are
we beginning to work out which are dangerous – only about one-
tenth of the substances in commercial use have ever been properly
tested. The well-known public disasters (Seveso, Flixborough,
New York's Love Canal area) represent only the tip of the iceberg
when it comes to controlling toxic chemicals. A report by Friends
of the Earth recently demonstrated that the uncontrolled use of
pesticides in Britain has become a major scandal. 'We reveal an
industry essentially out of control, with no public accountability,
yet with its toxic products invading our homes, air, water and
food.'[9] Internationally, the situation is even worse. Twenty-five
years after the publication of Rachel Carson's *Silent Spring*, more

than half of all nations do not have legislation controlling the use of pesticides. Conditions for farmworkers, already bad enough in the UK and USA, are appalling in the Third World, accounting for hundreds of deaths and thousands of illnesses. Many products that are banned in the home countries are then dumped in the Third World, invariably without adequate instructions. This sets up a 'circle of poison', which makes a nice irony for those intent on monitoring the pattern of suicidal industrialism, since much of the imported food from the Third World contains residues of those same banned pesticides. But this too, it seems, makes little impact.

In many countries, anti-pollution laws are under attack, accused of draining away scarce capital from more productive investment. This is demonstrable nonsense. The Organization for Economic Co-operation and Development (OECD), in its recent report, *Environment Policies for the 1980s*, commented: 'Environment policies are increasingly perceived to be justified on economic as well as social and ecological grounds. National studies, based now on several years of experience, suggest that total benefits are well in excess of the total costs of measures to abate pollution.'[10] For the uninitiated, this represents something of a breakthrough: good sense and ecological wisdom can even be justified economically! None the less, the emphasis is still on 'abatement', and this shows that we still have a long way to go. One of the largest growth industries in the USA is the 'pollution abatement' industry; now this may sound a good thing, but it means that even in 'curing' pollution, we are *causing* pollution through the further consumption of energy and raw materials. We're not likely to make much sense of all these problems till we start talking about prevention rather than abatement.

Permissible Poisons

In the case of the nuclear industry, prevention means getting rid of it *altogether*. Those who still talk of a 'safe' industry, or of technologies for 'permanently controlling' the dangers of plutonium and other nuclear waste products, have been sorely led astray by the so-called 'experts'. For it is in this area that scientific expertise has been most devalued. The debate over 'maximum permissible levels' of exposure to radiation has revealed countless examples of ignorance, short-sightedness and downright dishonesty. As the levels have been brought down lower and lower, the nuclear industry's own radiation experts find themselves supporting claims which only a few years ago they dismissed as the work of cranks and intemperate mischief-makers. And as we contemplate the dangers of carting spent nuclear fuel around all over the country, or of the disposal of nuclear waste through sea-dumping or in deep burial sites on land, one might well ask who exactly gives them permission to say what's permissible?

The most recent crisis yet again concerns the reprocessing plant at Windscale. (You can call it Sellafield, if you like: *plus ça change, plus ça reste la même chose*. It's still a dangerous, cancerous, murderous abomination.) In a searing documentary in November 1983, Yorkshire Television discovered that nearby villages had six times the national average cancer rate among children, and in the village of Seascale it was ten times the average. British Nuclear Fuels Limited (BNFL) refer to these cancer rates as 'chance happenings' and deny any connection between them and the plant. The National Radiological Protection Board (NRPB) has repeated time after time that radiation levels in the area are well below the 'permissible maximum'. Such authorities have completely lost credibility in the local area; people don't want to walk or play on the beaches; they even feel afraid in their houses, knowing that plutonium and other radioactive substances are accumulating in the very household dust.

The discharges from Windscale are by far the worst in the Western world, and it's now quite clear that the Cumbrian coastline has been seriously and permanently contaminated with lethal radioactive materials. BNFL has an appalling record, and there have been deceptions and cover-ups from the time the plant came into operation. The full details of the serious fire in 1957 have only just been revealed! Apart from major accidents, no account of unauthorized releases is ever given. Hence the importance of the events concerning Greenpeace in November 1983, when in an attempt to cap the discharge pipe they stumbled into a large radioactive slick with far higher levels of radiation than are offically 'permitted'. One wonders just how often this happens.

The decision by Greenpeace to take this kind of direct action was totally justified given the refusal of the government to take any action itself. They were not actually successful in capping the pipe, and were fined £50,000 for their troubles, a sum which might have driven them into bankruptcy had it not subsequently been reduced. But they were successful in arousing public concern and anger. I still smile in ironic delight at the memory of those officials having to cordon off an area in Whitehall after radioactive silt taken from the Ravenglass Estuary had been dumped there outside Downing Street. They may even have smiled in Ravenglass, though with some bitterness.

Acid from on High

The damage done by acid rain was first identified in Sweden in the 1960s, when the fish in the lakes started to die. By now, 18,000 lakes have been seriously acidified. Realization has dawned that acid rain is responsible for enormous amounts of damage, causing the slow acidification of the ground, interfering with the ability of trees and plants to absorb nutrients, and releasing toxic heavy metals, like aluminium and cadmium, into the soil and the water, poisoning both plants and fish. Buildings are being literally eaten away: the Acropolis has suffered more damage in the last twenty years than in the previous two thousand. The US Environmental Protection Agency estimates damage to buildings in the US from acid rain at more than $2 billion every year.

And all of this is caused by our industrial system. Not content with poisoning everything else, we've even managed to poison the rain. The pollutants concerned (sulphur dioxide and nitrogen oxides) are discharged into the air when fossil fuels are burned in power stations, factories or vehicles. Some of it falls near to its source; the rest reacts chemically in the atmosphere to form both sulphuric and nitric acids which can be transported thousands of miles before falling with rain or snow. Forty million tons of sulphur are pumped into the atmosphere over Europe every year, of which we contribute between 5 million and 6 million tons, more than any other country. Of this 2.6 million tons comes from our power stations, which makes the Central Electricity Generating Board (CEGB) the biggest single polluter in Europe. Up until now the most serious effects of this have been felt in other countries, but the UK is now beginning to suffer. South-west Scotland has been particularly affected by this English pollution, but in the Lake District and elsewhere the same warning signs as occurred in Sweden are now becoming apparent.

Building tall smokestacks to disperse the pollution is no remedy at all; it merely spreads it over a wider area. Putting lime in the acidified lakes is little better; as the Norwegians say, it's like taking aspirin to cure cancer. But emissions from power stations can be significantly reduced by a process called 'scrubbing', removing the sulphur from the emissions before they're released. It's expensive and will certainly lead to an increase in energy costs. A longer-term solution lies in the development of a new coal-burning technology called 'fluidized bed combustion', which not only gets rid of 90 per cent of sulphur dioxide, but reduces nitrogen oxides through burning the coal at lower temperatures.

The pro-nuclear lobby use the issue of acid rain to suggest that the development of nuclear power would be the cleanest and safest solution as far as the environment is concerned. Nothing is likely to make an ecologist hop with rage as fast as that particular idiocy, for it is precisely the *combination* of problems caused by acid rain and nuclear power that confirms the validity of two fundamental principles of the green alternative: that the conservation of energy is the best source of energy, and that *only* a more efficient, low-energy economy can possibly avoid the problems of fossil-fuel pollution.

The CEGB has responded to all this with its customary zeal. In September 1983 it announced a five-year, £5 million research project to investigate certain aspects of the problem. This is both typical and despicable: typical, because the CEGB has always demonstrated extreme unwillingness to acknowledge its responsibility, even denying that it has had anything to do with it until the evidence became irrefutable; despicable, since adequate research has already been done, and claiming that there is 'insufficient proof about certain details' is merely a cheap way of buying five years' respite. To make things even worse, the CEGB boss, Sir Walter Marshall, announced at the launch of the project that no action would be taken till the survey had been completed.

Contrast our efforts with those of Germany, which has now launched a major desulphurization programme, and announced stringent vehicle emission controls to be introduced in 1986. But in Germany the effects are there for all to see, for its forests are literally dying. One-third of Germany is forested, mostly with fir and spruce, and one-third of this is now affected by *Waldsterben* – forest death. It's all happened very suddenly and, ironically, at a time of recession when industrial pollution is reduced. But *Waldsterben* is the result of a gradual and progressive chemical onslaught that weakened the trees for many years before they started dying. The resulting losses will be on a par with the destruction of the tropical forests, both in different ways caused by the very process of industrialization. In a startling and dramatic fashion, some of the largely *hidden* costs of industrialism have suddenly become visible.

Global Warming Warning

The least precise and tangible of all the many threats to the environment concerns the build-up of carbon dioxide in the atmosphere. We are engaged in a dangerous planetary experiment. When fossil fuels are burned, carbon dioxide is released into the atmosphere. There's no technological fix that can do anything about this, and at the present rates of consumption, CO_2 concentrations will have doubled by the middle of the next century. This may well trigger off the 'greenhouse effect', causing significant and socially traumatic climate changes. An increase of 2 °C would be enough to melt much of the Antarctic.

The threat of a 'global warming' confirms much of what I've been saying throughout this chapter. By endlessly seeking solutions through *more* of this or *more* of that, we are merely compounding the problems. We need to think in terms of *less* being more efficiently used. Anything else will continue to foster the illusions of contemporary industrialists and technological fixers. As Schumacher so cogently explained, their illusions are based on a failure to distinguish between income and capital: 'Every economist and businessman is familiar with the distinction, and applies it conscientiously and with considerable subtlety to all economic affairs – except where it really matters: namely, the irreplaceable capital which man has not made, but simply found, and without which he can do nothing.'[11]

That then is the framework: an environment ravaged and devastated by a majority of the world's people just so they can live from day to day, and by a minority to satisfy often wasteful and mindless habits of consumption. The poor haven't the luxury and the rich haven't the inclination to think about tomorrow. As the pressures become more obvious, one might suppose that more sustainable patterns of economic development would be introduced as quickly as possible. But more often than not the so-called 'imperatives' of our industrial culture, combined with the indifference, ignorance and lack of vision of most politicians, ensure that ecological concerns remain firmly at the bottom of any Government's list of priorities. We believe these concerns are the *number one* priority – and that's where the politics of industrialism and the politics of ecology come into direct conflict.

4
Industrialism in All its Glory

Now That The Twain Have Met...

The claim made by green politics that it's 'neither right, nor left, nor in the centre' has understandably caused a lot of confusion! For people who are accustomed to thinking of politics exclusively in terms of the left/right polarity, green politics has to fit in somewhere. And if it doesn't, then it must be made to.

But it's really not that difficult. We profoundly disagree with the politics of the right and its underlying ideology of capitalism; we profoundly disagree with the politics of the left and its adherence, in varying degrees, to the ideology of communism. That leaves us little choice but to disagree, perhaps less profoundly, with the politics of the centre and its ideological pot-pourri of socialized capitalism. The politics of the Industrial Age, left, right and centre, is like a three-lane motorway, with different vehicles in different lanes, but *all* heading in the same direction. Greens feel it is the very direction that is wrong, rather than the choice of any one lane in preference to the others. It is our perception that the motorway of industrialism inevitably leads to the abyss – hence our decision to get off it, and seek an entirely different direction.

Yet it's built into our understanding of politics today that capitalism and communism represent the two extremes of a political spectrum. The two poles are apparently separated by such irreconcilable differences that there is no chance of them ever coming together. According to such a view, the history of the world from now on (however long or short a time-span that may be) is predicated upon the separateness of these two ideologies.

There are, indeed, many differences; in social and political organization; in democratic or totalitarian responses; in economic theory and practice. But for the moment, let's not dwell on these. Let us consider the *similarities* rather than the differences. Both are dedicated to industrial growth, to the expansion of the means of production, to a materialist ethic as the best means of meeting people's needs, and to unimpeded technological development. Both rely on increasing centralization and large-scale bureaucratic control and co-ordination. From a viewpoint of narrow scientific rationalism, both insist that the planet is there to be conquered, that big is self-evidently beautiful, and that what cannot be measured is of no importance. Economics dominates; art, morals and social values are all relegated to a dependent status.

I shall be arguing two things in this chapter: first, that the similarities between these two dominant ideologies are of greater significance than their differences, and that the dialectic between them is therefore largely superficial. If this is the case, it may be claimed that they are united in one, all-embracing 'super-ideology', which, for the sake of convenience, I intend to call industrialism. Secondly, that this super-ideology, in that it is conditioned to thrive on the ruthless exploitation of both people and planet, is *itself* the greatest threat we face. As Roszak puts it: 'The two ideological camps of the world go at one another; but, like antagonists in a nightmare, their embattled forms fuse into one monstrous shape, a single force of destruction threatening every assertion of personal rights that falls across the path of their struggle.'[1]

If that is so, there must be something with which we can replace it; not another super-ideology (for ideologies are themselves part of the problem), *but a different world view*. That is the not unambitious role that green politics is in the process of carving out for itself.

Tweedledum and Tweedledee

For an ecologist, the debate between the protagonists of capital-ism and communism is about as uplifting as the dialogue between Tweedledum and Tweedledee. That most commentators and politicians still consider it to the the be-all and end-all of politics only serves to demonstrate the abiding delusions of our industrial 'wonderland'. Consider it, however, we must. So let us start with one of the central tenets of capitalism: that by continuously expanding the production of goods and services, and simultaneously promoting their consumption, gainful employment will be found for all, and enough wealth will be created to meet society's needs. How that is achieved is a matter for different Governments, but the general idea is that such wealth will 'trickle down' from top to bottom.

Every time I hear that phrase, 'trickle down', I'm reminded of those nineteenth-century cartoons of bloated industrialists at groaning tables, usually with gravy dribbling down their chins and falling in thick gobbets on to starched table napkins, while drawn and emaciated workers look on and fill their bellies with dreams. After a long and bitter struggle, something has indeed trickled down, and the standard of living for millions in the developed world has improved accordingly. But that's where the trickle dried up, still with grotesque disparities existing between the richest and the poorest.

Over the last fifteen years or so many studies have shown that these disparities are *not* shrinking, and in some cases are getting worse. This is the case both within individual nations and in terms of an international distribution of wealth. In many respects, economic growth has served as a substitute for equality of income: as long as everybody was getting something, large income differentials remained acceptable. This worked well, just so long as energy and resource inputs remained cheap and dependable. But if economic growth has to be stabilized *at some point* (and all can surely agree on that without quibbling about when), there can be no conceivable moral justification for relying on 'trickle-down' to help the poorest. Resources will give out long before such an uneven distribution of wealth could possibly provide for those at the bottom of the heap.

Over and above this basic failure of capitalism, there is a far more insidious contradiction in the way the system seeks to meet people's needs. At the heart of the problem is the fact that wealth is basically distributed through the jobs people do. The only reason why many people do the jobs they do is to earn an income.

As well as producing the things we do need, this means that many things are produced which we don't need. People produce junk and cajole others into buying it, purely as a means of earning an income. If the system works (i.e. we achieve full employment), we basically destroy the planet; if it doesn't (i.e. we end up with mass unemployment), we destroy the lives of millions of people.

Yet economists continue to neglect the essentials involved; ecological degradation is dismissed as a 'negative but controllable externality'. Such complacency persists despite the fact that in the last decade the relationship between expanding economies and the biological systems that underpin these economies has changed drastically. In 1972, massive purchases of wheat by the USSR doubled the world price in months. In 1973, OPEC quadrupled the price of oil. Both were blamed for what they did, but the point is that these were only symptoms of a global problem, not the causes. We must return to the stark reality of our population/ resource ratio: by the time the world population reached 4 billion in 1976, per capita production of oil and many other commodities had *already* peaked.

Even now few people understand the impact that cheap oil had on the world economy. Together with the demand-stimulation of Keynesian economics, it fuelled an extraordinary increase in prosperity and trade. In *Building a Sustainable Society*, Lester Brown demonstrates the many ways in which cheap oil radically transformed the global economy through the build-up of an international transport network, the vigorous export expansion programmes of most countries, and the evolution of car-domin-ated transportation systems. Above all, cheap oil revolutionized agriculture, making it possible to achieve astonishing gains in food output. Chemical fertilizers and pesticides were the key to this, removing many natural constraints on food production.

Cheap oil pushed back resource constraints in every continent. But by the same token, as we've already seen, when oil reserves start to dwindle, this process will be reversed and increasing pressure will be put on basic biological systems. There will be a similar downturn in manufacturing and industry, for oil is still the key input into much of world industrial production. The evidence is already mounting: between 1950 and 1973 the world economy expanded at a record 4 per cent per annum, between

1979 and 1982 by just 1.6 per cent per annum. That alone should be enough to remind us that continuous rapid economic growth was *never* a permanent feature of the system, *never* a part of the natural order.

The Law of More and More

But that's exactly what most people, and certainly most politicians, do believe. And in order to achieve growth, more things must be produced. And for more things to be produced, more things must be consumed. The more people consume, the better it is. The shorter the life a product has, the better it is. It's not so much a question of consumer durables as of durable consumers. And in order to achieve this, consumers must be manipulated into the smoothest possible cycle of acquisition and disposal, into a uniform, superficial understanding of personal and social requirements. Consumption becomes an end in itself. Even when the market reaches saturation, the process doesn't stop; for the only way to beat a glut is to turn everybody into gluttons. The consumer rules, we are told, but as the gilded web of materialism tightens around us, who rules the consumer?

It was the fiery analysis of Narindar Singh's extraordinary but little read book, *Economics and the Crisis of Ecology*, that first opened my eyes to what he calls this 'embrace of death between mass-production industrialism and our mass-consumption society.'[2] Capitalism depends on demand stimulation; in order to prosper it has to create new demands that it then seeks to satisfy so as to forestall the threat of mass unemployment. When it comes down to it, the throw-away economy is a pretty efficient means of achieving just that. But the logic of ecology stands in direct opposition to the logic of industrialism; for it is clear that in the very process of 'succeeding', industry cannot help but destroy its own material base. High consumption simply cannot be indefinitely sustained on a finite planet.

We are thus trapped on a treadmill by the very logic of industrialism. In order to avoid mass unemployment, ever higher levels of consumption must be stimulated; but the increases in GNP which then result can only be achieved through the more rapid depletion of exhaustible resources and the deterioration of our environment. We're right back where we started, with our two competing world views, and the different perspectives of rationality those two views provide. From an industrial point of view, it's rational to behave in this way, to promote wasteful consumption, to discount social costs, to destroy the environment. From a green point of view, it's totally irrational, simply because we hold true to the most important political reality of all: that *all* wealth ultimately derives from the finite resources of our planet.

Confirmation of these irreconcilable perspectives on rationality was recently provided for me in a debate at Durham University with one of this country's leading industrialists. Having grudgingly conceded that there were indeed constraints on many of the world's resources, he used this to persuade his audience of undergraduates that 'they should get out there and compete for these resources with all the more determination *while they're still there.*' That says it all really: capitalism as we know it simply cannot provide the preconditions for an ecologically sane, humane economy.

No more can communism. Irrespective of who owns the means of production, the unrelenting pursuit of growth and industrial expansion must necessarily degrade the planet and impoverish its people. Socialization of the means of production makes little difference: what are vices under capitalism do not become virtues under communism. A filthy smokestack is still a filthy smokestack whether it is owned by the state or by a private corporation. Indeed, Boris Komarov's book *The Destruction of Nature in the Soviet Union* shows that managers and industrialists in the USSR often don't have to bother even about the inadequate controls that exist in Western democracies.

From an ecological point of view, communism is merely an extension of capitalism by other means. In determining to 'overtake' capitalism, communism must first follow it, with a vision of progress based on the same methods of production, the same division of labour, and the same materialist consensus. Lenin once wrote, 'Communism is Soviet power plus the electrification of the whole country.' Some vision. It is true, of course, that communist countries are not as obsessively consumer-oriented, but by all reports it isn't for want of trying. There are certainly very steep differentials in wealth and privileges in most communist countries. Whether it is organized through the Market or according to the Plan, the maximization of production and consumption produces the same irrational results. And because all socialist countries are meshed into the system of world trade just as inextricably as capitalist countries, none has ever come near the ideal of producing only for *need* rather than for profit or exchange. In this sense at least, the two extremes are not really extremes at all, but merely two ways of designating the same thing. Rivals no more, united in their industrialist super-ideology, the nations of East and West blithely go about their business of destroying the planet.

There are some tough lessons to be learned here. Environmentalists may, for instance, find it difficult to understand that there are limitations to the limits to growth argument. It's all very well demonstrating these limits to growth, but it's less convenient to have to acknowledge that industrialism cannot possibly help violating them, given its inherent need to expand and the toxic nature of much of its output. Curbing the side-effects fails to tackle the problem at source, and may even shift the emphasis on to pseudo-problems. Grandiose schemes to recycle everything may conceivably be desirable, in that curing an illness is better than promoting it; but as we all know, prevention is better than cure. The radicalism that informs the politics of ecology leads to tough but always logical conclusions. Sustainablility and industrialism are mutually exclusive. As Narindar Singh says, 'The prevailing order cannot solve the problems of its own creation. It can only intensify them relentlessly.'[3] This becomes abundantly obvious when we look in a little more detail at the role of technology in society today, for technology is one of the most important of the many pressures promoting inherently unsustainable patterns of production and consumption.

Whose Choice?

For most people there's no ambivalence about technology. It's a good thing. There's a sort of general consensus that technology will take us wherever we want to go, and get us out of whatever mess we're in. The more complex the 'fix', the better it is. This adulation crosses all political boundaries. To be opposed to current trends in technological development is to put oneself beyond the industrial pale, to be branded as a Luddite (more of whom later), a backwoodsman, a Neanderthal. Greens find this really rather trying, for we think it's about time a few more questions *were* asked, not least about the element of 'technological determinism' (i.e. what can be done, will be done) that seems to be taking over. Decisions are shaped not by traditional political processes, but by available technological capacity. Lobby groups work hard to tell us that 'we have no choice', and since it's all in our interest anyway, why worry?

Fortunately more and more people are worrying, for modern science and technology are themselves major elements in the ideology of industrialism. There are those who would still have us believe that science itself is neutral, yet more and more it is being put to ideological uses to support particular interests, especially by those who already wield the power in our society. Science is simply not geared up to cope with the priority problems of humanity. It is the already privileged sectors of the developed economies that seem to get most of the benefits, spurred on by those whose interests can hardly be described as neutral. These 'technocrats' have ensured that the principal measure of civilization should be technological progress rather than wisdom, compassion or mutual co-operation. So the race for space goes on even as thirty children die of starvation every minute. Giant, inappropriate, destructive technology still wins the lion's share of new investment. Concorde rules, OK! And as we shall see, science plays a major role in the promotion of militarism in all industrial nations.

We are left to look on, like so many sorcerers' apprentices, as the technological broomsticks take over. This process has a momentum all of its own, and whether or not we can slow down that momentum has become a matter of major political significance. In the *Pentagon of Power*, Lewis Mumford wrote:

Western society has accepted as unquestionable a technological imperative that is quite as arbitrary as the most primitive taboo:

not merely the duty to foster invention and constantly to create technological novelties, but equally the duty to surrender to these novelties unconditionally, just because they are offered, without respect to their human consequences.[4]

Atoms for War and Peace

It's the human consequences of much of modern technology that so concern many green activists. The difference between 'Atoms for Peace' and 'Atoms for War' provides a classic case of modern man being unable to think further ahead than the end of the next financial year. The US 'Atoms for Peace' programme was the initial vehicle for spreading nuclear technology. Codenamed 'Operation Wheaties' (after President Eisenhower's breakfast cereal!), the programme was given the usual hard sell, complete with subsidies and training in nuclear technology. More than thirty years later, we are confronted with the apparently inexorable proliferation of nuclear weapons as one country after another uses atoms for peace to manufacture atoms for war.

In addition to the USA, USSR, UK, France and China, three other countries are believed to have nuclear weapons: Israel, South Africa (whose first research reactor was supplied by the USA) and India (which gained its expertise from US and Canadian nuclear scientists). The continuing export of nuclear technology ensures the future production of nuclear weapons in many other countries. There is the terrifying possibility of new nuclear blocs: the Islamic bomb (linking Pakistan, Libya and Iraq), or the South Atlantic bomb (Argentina and South Africa). And there's a long list of twenty or more countries that might develop the technology at some stage in the future. Many have refused to sign the Non-Proliferation Treaty, which forbids the development of nuclear weapons, or even to put their nuclear reactors under international supervision – either refusal is tantamount to a declaration to develop weapons at some stage.

A reactor programme is all a country needs to start work on a weapons programme. Reactor-grade plutonium, the by-product of nuclear fission, *can* be used for weapons, and in 1977 the USA successfully tested such a device. The competition to sell reactors overseas remains fierce; the 'if I don't, someone else will' mentality dominates any debate about the desirability of promoting further proliferation, and the safeguards surrounding such sales are often totally disregarded.

So the genie is well and truly out of this particular bottle, and the attitude of successive UK Governments has done little to impede its progress. Our own nuclear reactors were a direct spin-off from an initial nuclear weapons programme based on the purpose-built reactors at Calder Hall and Chapelcross. That early connection has been maintained, though it's now so shrouded in secrecy that it's well-nigh impossible to discover the truth. It's not even certain how much plutonium has been produced from our nuclear reactors: it could be as much as 40 tons or as little as 28 tons. It's known that there are 12 tons in civil stock, that 6 tons have been used for research on the Fast Breeder Reactor (FBR), and that 1.3 tons have been sold abroad for research. The rest is either being kept for the FBR programme (which, for various reasons, is most improbable), or it has been sold, possibly to France, or more probably to the USA in exchange for supplies of enriched uranium of a higher quality than we otherwise have access to. Whether it's then used in the USA for weapons, rather than for research, is academic: either it isn't, in which case US supplies are freed for that purpose, or it is.

But the basic problem remains unsolved. Is plutonium from our reactors being used to make nuclear weapons in *this* country? At the Sizewell Enquiry, John ('Stonewall') Baker, the CEGB's chief policy witness, claimed that 'no plutonium produced in CEGB reactors has ever been applied to weapons use either in the UK or elsewhere.' Clear enough, but all the evidence points to exactly the opposite! Some of our nuclear weapons are made in the USA, but most are made here. Some of the plutonium for them comes from Calder Hall and Chapelcross, and some is recycled from old warheads. But those two sources provide nothing like enough plutonium for *existing* nuclear weapon stocks, let alone for the proposed expansion required for Trident and the Tornado.

Further information is apparently not available for 'national security reasons'. But if no reactor plutonium has ever been used for weapons, what has national security got to do with it? Sadly, it has become more or less futile to ask such questions. The 'technological imperative' that drives our industrial society allows for no check or restraint along the suicidal path we have chosen to follow – or rather, are informed we have no option but to follow. The forces of contemporary industrialism impose so uneasy a consensus upon us all, and impose it now with such rigour, that like the pigs and the humans at the end of Orwell's *Animal Farm*, the leaders of the capitalist world and the leaders of the communist world have become all but indistinguishable. And they depend upon us, the farmyard animals, not realizing what's going on – for how else could they keep us in subservience?

5
The World at War

Who's Afraid of the Big Bad Bear?

So much for the similarities between capitalism and communism. Let us return to the differences, for despite sharing with the West in what I have called 'the super-ideology of industrialism', despite significant trade between East and West, despite continuing social and cultural contacts, the USSR remains the implacable enemy, personifying in the public mind all that is wicked. We are led to believe that the so-called 'irreconcilable antagonisms' between the two political and social systems promote in each the desire to vanquish the other.

As anyone who has been out canvassing for CND and other peace campaigns can testify with some feeling, the perceived Russian threat remains the greatest obstacle to nuclear disarmament in this country. Many is the time, when all other arguments are won, that the vision of the Russian hordes storming over the horizon takes one right back to square one. The key question for the peace movement is to work out whether this is an objective assessment, or misplaced hysteria being whipped up for different reasons. Given the high stakes and the subtle use of propaganda on both sides, one has to tread with caution here. I have relied heavily on Jim Garrison's authoritative analysis in *The Russian Threat*, which I would reckon to be by far the best book on the whole subject.

Even the most fleeting of glimpses at Russian history makes one aware of the sense of vulnerability that the Soviet Union must feel. Continuing encirclement, repeated invasions (most recently by the Nazis, which claimed the lives of 20 million Russians), and a feeling of isolation must obviously shape its thinking. Through massive military expenditure it has built up a most effective war machine, often in the old-fashioned belief that more of something necessarily means better. It 'over-insures' against the fear of war which it believes the West, or China, seeks to wage against it. Its role in Eastern Europe should be understood in these terms. The Warsaw Pact countries are its buffer zone, and the USSR is determined they should remain that way. The primary function of the Pact is therefore to deny to the countries of Eastern Europe control over their own military and political structures. It justifies this imposed control on the basis of defending those countries against aggression from the West. The current build-up of NATO forces merely provides the necessary pretext to reinforce Soviet domination within the Warsaw Pact, and for this it must be grateful. There's nothing it fears more than the collapse of its influence in this area, for that would mean the collapse of its buffer zone. The events in Poland demonstrate yet again how crucial this is for the USSR, for if its 'allies' start opting for the West, what then becomes of its concept of national security?

Its presence in Afghanistan is not as easy to account for. It was presented in the West as part of a grand design to seize the oil fields of the Persian Gulf, but the likelihood is that the Soviet Union acted *defensively* for political reasons rather than *offensively* for military reasons. The politics of Afghanistan are horrendously complicated at the best of times, but when the traditionally non-aligned monarchy was overthrown in 1973, things became chaotic. The USSR was faced not only with the prospect of a more pro-Western country on its borders, but also with a very real threat to the stability of its own Moslem people. In December 1979 it marched in with 85,000 troops, the first time since 1945 that it has moved beyond the boundaries of the Warsaw Pact in hostile action.

It was, of course, a terrible mistake. (Afghanistan is a fiercely nationalistic country, and having defeated the British Empire in the eighteenth and nineteenth centuries, it will no doubt eventually defeat the Russians.) But it's absurd to make out that the USSR invaded so as to threaten the Persian Gulf. It is already the world's largest producer of oil, with massive deposits in Siberia still to be exploited. Such a suggestion is a classic example of the prevailing paranoia; there is indeed no possible excuse for the Russian invasion of Afghanistan, but the reaction from the West was wholly misguided. Such paranoia stems from the lurking belief that the USSR is gradually building up military superiority so as to take over the world.

In the age of nuclear weapons, even to talk about 'superiority' becomes an exercise in profound irrationality. The 'overkill' capacity is such that the USA could destroy all major Soviet cities about forty times over, and the USSR could do the same to America about twenty-five times over. None the less, the numbers game remains at the heart of the disarmament process. At the strategic level, the USA has more warheads and they're more accurate. The USSR's warheads are bigger and more powerful, and it has more delivery vehicles. In terms of intermediate nuclear weapons, the Soviet Union has a slight advantage in numbers, but not in capability. With the short-range theatre weapons, NATO forces have very considerable superiority. As regards conventional weaponry, NATO has more or less the same number of troops as the Warsaw Pact; moreover, the quality and efficiency of NATO weapons more than compensates for numerical inferiority in certain categories. Finally, as regards annual military budgets, it is clear that NATO countries consistently and considerably outspend Warsaw Pact countries.

The idea of the USSR bent on world domination is simply not credible. Surrounded by enemies, outspent, outnumbered, outgunned, technologically and economically inferior, there is absolutely no evidence either that it wants war or that it plans to invade. There has been a clear weakening of influence in the Warsaw Pact countries, it's lost control of the Communist Parties in Western Europe, it has a disaster on its hands in Afghanistan, it's been thrown out of many Third World countries, and remains in just a few relatively unimportant little nations: the Russians are basically lousy imperialists. To cap it all, the drain on their economy is appalling; high expenditure on arms damages them more than the USA, and Cuba alone is costing them $6 million every day. They desperately need the trade with the West. The Soviet Union should surely be credited with an understanding of the limits to its power.

The notion of world domination by any single power is, anyway, a thing of the past. The Soviet Union certainly has formidable capabilities, a ruthless and utterly abhorrent internal system, an appalling record on human rights, and may indeed want to spread communism around the world. But to combat that, we'd be better off investing in bread and Bibles than nuclear weapons.

The Cold War, Part Two

The facts then are very much at odds with popular perception. But it's the latter which counts, for it's on the basis of what ordinary people are supposed to think that the two rival alliances confirm and strengthen their adversary postures. Like it or not, the Cold War is with us again. In his pamphlet *Beyond the Cold War* E. P. Thompson explains how this has come about:

> The Cold War may be seen as a show which was put upon the road by two rival entrepreneurs in 1946 or 1947. The show has grown bigger and bigger; the entrepreneurs have lost control of it, as it has thrown up its own managers, administrators, producers and a huge supporting cast; these have a direct interest in its continuance, in its enlargement. Whatever happens, the show must go on.'[1]

That is to say, the Cold War is basically about itself. It has taken on a momentum all of its own; it is entirely self-reproducing. For the show to go on, there must of course be two adversaries, so that the hawks of NATO can feed off the hawks of the Warsaw Pact, and so on. Such a suggestion is not entirely irrational, for in a very interesting passage, E. P. Thompson goes on to consider the extent to which societies have always been bonded together by a threat from outside. Historically, the threat of an enemy, of the 'Other', has always provided a pretty useful means of reinforcing social discipline and cohesion; the greater the threat, the greater the bonding. The USA, with its fragmented, 'melting-pot' society, can ensure social discipline only as the 'leader of the Free World'. The Soviet Union, with its ramshackle empire of very different cultures, achieves the same only as the 'heartland of socialism'. In the West, dissidents are successfully relegated to the margins of politics by such a process; in the East they are ruthlessly persecuted and suppressed.

What makes the Cold War *entirely* irrational is that it has now become a permanent and necessary part of the internal politics of both sides. Moreover, such Cold War posturing, which could, and in the past often did, end in actual war, is still being acted out on the basis of the rival ideologies of the 1940s. With our present weapons this can only lead to disaster, for as the military establishments grow, the adversary postures become more irrational. The Church of England's report, *The Church and the Bomb*, put it this way: 'Mutually stimulated paranoia is blinding all concerned to the way their opponent is likely to behave. The prophecies of aggression are self-fulfilling prophecies.'[2]

The 'Holocaust Lobby'

So who then are the participants in this drama currently being acted out in the theatre of Europe – the directors, producers and cast? Though some prefer to stay in the wings, the others are easily recognizable, and if I base my review on what happens in the West, that's because I happen to be watching the show in the West, not because I think the Russians are necessarily any better. The whole point is that one side is the mirror image of the other.

The first two groups involved are the scientists and what is referred to as the 'military-industrial complex', that cosy cluster of relationships between the armed forces and big business. And the weapons business is big, big business. Every year the Pentagon places orders worth around $80 billion. But the odd thing is that most of the weapons produced are never actually used in war, with the result that those corporations most involved become dangerously introverted: the driving force behind them is not the defeat of any enemy, but that of an endless sophistication of the art of war. As soon as work is finished on one weapons system, work starts on another, and for those businesses which have become dependent on defence contracts, the incentive to get new contracts is compelling.

Nowhere is the power of the 'technocracy' more apparent than in the world of arms production. Often it is the researchers in the laboratories who propose new schemes or 'improvements', not the generals or the politicians. Lord Zuckerman, a former Chief Scientist to the British Government, has written: 'It is he, the technician, not the commander in the field, who starts the process of formulating the so-called military need. It is he who has succeeded over the years in equating, and so confusing, nuclear destructive power with military strength', with the result that 'the men in the nuclear weapons laboratories of both sides have succeeded in creating a world with an irrational foundation, on which a new set of political realities has in turn to be built.'[3]

As a classic example of technological determinism, the solutions often emerge from the scientist or the corporations, before anybody realizes there's a problem! Control of such a complex is almost impossible, as President Eisenhower realized all too clearly in his last speech as President: 'We have been compelled to create a permanent armaments industry of vast proportions... We must not fail to comprehend its grave implications. In councils of government we must guard against the acquisition of unwarranted influence, whether sought or unsought, by the military-industrial complex.'

Easier said than done, especially when the politicians, the third main group involved in the Cold War drama, are only too keen to consolidate this 'unwarranted influence'. The present American administration works hand in glove with the military-industrial complex. Ronald Reagan was elected to the White House at the peak of the latest anti-Soviet wave of feeling in the USA. What to others has been merely anti-Soviet rhetoric, to Ronald Reagan is gospel. His administration is loaded with Cold War warriors, none more belligerent than Paul Nitze, who headed the US delegation at the Geneva disarmament talks. Since 1956 Nitze has been an open advocate of a 'war-winning strategy'. In an article in that year, he wrote: 'It is quite possible in a nuclear war that one side or the other could win decisively... the greater our superiority, the greater are our chances of seeing to it that nuclear war, if it comes, is fought rationally.' (I shall not labour the point about that particular use of rationality.)

Winnable, rational nuclear war was a strategy totally at odds with the declared doctrine of Mutually Assured Destruction (MAD). It was the very fact that nuclear war was thought to be unwinnable that made deterrence a viable strategy. Nitze's position amounts to a refusal to believe that the existence of nuclear weapons has in any way altered the nature of armed conflict. He has not changed his views since the 1950s, but whereas he once was a part of a tiny minority, he is now part of a majority within the current administration. In 1976 he was involved in setting up the Committee on the Present Danger (CPD), which aimed to reshape American foreign policy. Among its members was Ronald Reagan.

After his election in 1980, Reagan installed in his administration no fewer than thirty-two CPD members – familiarly known around Washington as the 'Holocaust Lobby'. For a while, Nitze became chief Strategic Arms Limitation Talks (SALT) negotiator (an appointment that historian Barbara Tuchman likened to 'putting Pope John Paul II in charge of abortion rights'), but SALT II was quickly ditched. The idea of the USA achieving nuclear superiority over the USSR was formally acknowledged in April 1981, when Caspar Weinberger, the Secretary of Defense, announced contingency plans to fight a 'protracted nuclear war'.

One of Weinberger's colleagues is a man called T. K. Jones, who became Deputy Under-Secretary of Defense. He features prominently in Ronald Scheer's extraordinary book *With Enough Shovels*, a title he took from one of 'TK's' more astounding comments. He suggested that in the event of a nuclear war you should 'dig a hole, cover it with a couple of doors, and throw three feet of dirt on top... it's the dirt that does it... if there are enough shovels to go around, everybody's going to make it.'[4] It's enough to make you weep – quite literally. Scheer's book is an eye-opener if you still believe, despite everything you hear to the contrary, that things are being conducted rationally out there. Paul Warnke (Jimmy Carter's chief SALT negotiator) has commented: 'The idea of winning a protracted nuclear war is absolutely crazy. For the first time this utterly nutty idea is clearly spelled out as the centerpiece of American defense strategy.'

I don't know about you, but that terrifies me. It becomes clearer and clearer that nuclear weapons are not just a nasty mistake in an otherwise healthy world. They are the logical outcome of the kind of society we have created for ourselves, the epitome of an exploitative, uncaring, unthinking world view. Our dependence upon such weapons has exposed a deep-rooted social sickness, of which the Cold War is just one chilling symptom. Jim Garrison sums it all up:

> We have made the Cold War an addiction. The scientists design the weapons for it; the corporations produce and profit from it; the military promotes and grows from it; the politicians rationalize it; the strategists plan for it. The military-industrial complex that has arisen *because* of the Cold War now *generates* the Cold War. It does not serve the interests of the people; people now serve it. And yet it is all of our own making.[5]

The UK Arms Blues

The UK is right in there doing its bit for the Cold War. Our Government is now spending about £16 billion a year on defence, £5 a week for every man, woman and child. That amounts to 6 per cent of GNP, an allocation which has grown by a quarter in real terms since 1978, far more than that of any other industrialized country apart from the USA. More than 10,000 companies have defence contracts; not including the armed forces, 700,000 workers are employed in defence-related work. The Government goes to great lengths to explain that military spending is important not only for defence purposes, but also as an instrument of economic policy. As such, it has been an unmitigated disaster, weakening the more productive sectors of domestic economy through lower rates of investment, lower productivity and fewer jobs. Fifty-four per cent of government-sponsored Research and Development (R & D) is spent on arms and defence, compared with 37 per cent in France, 10 per cent in West Germany and 4 per cent in Japan. Despite various so-called 'spin-offs' (from the biro to the nuclear reactor), it's complete gobbledegook to claim that we benefit more from such indirect technological advances than we would from direct investment in socially useful R & D. Moreover, spin-offs from the arms business are becoming less and less frequent. Military production is now mostly concerned with product improvement, and has resulted in an extravagant technology of very little relevance to anyone outside the specialized world of weapons. By civilian standards, military production is often hopelessly inefficient. It's also a downright lie to claim that more jobs are created by investment in the arms business than by equivalent investment in other areas, such as housing, education or health care.

Many people who work in the arms industry feel understandably threatened. Thousands of jobs are already disappearing in the industry, and it goes without saying that nothing is being done to find alternatives. The fate of Barrow highlights this problem, totally dependent as it is on the Vickers shipyards and works, in an area where there's already 11 per cent unemployment. When it was announced that the Trident submarines were to be built there, it seemed to guarantee some sort of job security. But although Trident is to cost an absolute minimum of £10 billion, it is typical of the trend in modern arms manufacture towards capital-intensive rather than labour-intensive spending, and many jobs will almost certainly be lost.

The huge overheads involved in arms production mean that businesses must try to develop an export market, which takes us right back to the old problem of first creating the wants that then have to be satisfied. In 1983, a worldwide total of more than $600,000 million was spent on arms at the rate of $1 million every minute. A significant proportion of this trade provided weapons for military dictatorships, which then used them for external aggression or internal repression. The endless spiral of expenditure on arms ensures that the Third World stays permanently poor. The money spent on weapons is priceless foreign exchange, either borrowed or earned through exports. Either way, the poor suffer. Boosting exports usually means pushing peasants off their land so as to make it available for cash crops for export to pampered consumers in the West. Borrowing more money increases the debt burden, the interest on which goes to Western banks rather than to the poor. The inevitable result is poverty, opposition and conflict, at which stage the arms come in very handy to suppress dissent in the name of stability and 'freedom'. And so the cycle goes on.

Since 1960, at least 10,700,000 people have died in sixty-five wars fought on the territory of forty-nine countries – a rate of 1,330 dead for every day of the past twenty-two years.[6] The dreadful figures detailing the damage caused by militarism go on and on: 500,000 scientists, a quarter of the world's total, are involved in military research; two governments in three now spend more on arms than on health, etc., etc. The statistics just seem to bounce off people nowadays, as if such problems were so far beyond the grasp of rational people as to justify an apparent lack of concern.

The diseconomies of scale have had as significant an impact on the imagination as they have on the environment; the more noughts there are, the more our minds are numbed. The morally abhorrent and ecologically suicidal are readily taken for granted, absorbed with that same fatalistic ease with which this country now tolerates startling symptoms of social and economic breakdown, not least the devastating burden of mass unemployment.

6

The Collapse of Economics

One Billion Jobs

The scale of unemployment in Britain today is by far the most serious problem this country faces. Yet you'd be a brave person or a desperate politician to suppose that it's ever going to get any better. Since the mid-1960s, there has been a clear tendency for unemployment to rise independently of the general level of demand, a trend which will be made much worse by the accelerating introduction of the new technology. Such unemployment is indeed 'structural'; it's become a permanent part of the structure of our industrial society.

The cost of existing levels of unemployment is already having a disastrous effect on the economy, amounting to more than £8 billion per annum in lost taxes and unemployment benefits. This figure doesn't even begin to measure the hidden social and personal costs of unemployment. Unemployment still carries a considerable stigma, and several research studies confirm that there is a high correlation between unemployment and rates of early mortality, disease, alcoholism, mental illness and crime. Not only does this cause untold personal suffering, but at a time when other resources are becoming scarce, it represents a serious and totally irrational waste of our most valuable productive resource. The wastage is bound to become more pronounced, since we still face a temporarily expanding labour market on account of the age structure of the population and a number of other social factors.

The global problem dwarfs even this appalling outlook. Between now and the year 2000, the workforce in developed nations will expand by 6 million a year, but in underdeveloped nations by a staggering 31 million a year.[1] Added to the existing numbers of unemployed and underemployed, it means we are talking of finding approximately 1 billion jobs by the year 2000. And yet, in all seriousness, politicians of every political complexion still suggest that the only possible remedy is to 'grow our way out of trouble'. It remains the established faith that once the recession is over, we shall be able to achieve sufficiently high rates of growth to ensure that the global economy flourishes as never before.

The relationship between economic growth and increased productivity is central to an understanding of unemployment. Forecasts of the levels of growth needed to reduce unemployment are largely unreliable, to the extent that they tend to underestimate technological change. This wilful refusal to face the facts is usually to be found doing the political rounds in the form of two ever-popular fallacies. The first of these is the fallacy that as long as we adopt new technologies as fast and as efficiently as our competitors, reinvestment in industry will generate significant numbers of new jobs. In fact, if one looks at agriculture, car manufacturing, the petrochemical industry, and many other large-scale sectors of the economy, employment almost always tends to go *down* as output goes up. The level of demand is an almost irrelevant factor: up or down, the number of people employed varies very little. Private investment to maximize production and profits necessarily requires a reduction of labour costs, and extensive application of the new technology, *whatever* the nominal levels of growth to which it contributes, will cause the loss of millions of jobs.

The second fallacy is that the service sector will continue to expand, as it has done in the last decade or so, to take up the slack. In the first place, the service sector in the UK partly depends on a healthy manufacturing sector, and it is clear that, if anything, the impact of micro-technology will be greater in the service sector than anywhere else. Moreover, as Jay Gershuny, of Sussex University's Science Policy Research Unit, pointed out in his book *After Industrial Society: The Emerging Self-Service Economy*, 'Services which were previously provided from *outside* the household, are now replaced by production *within* the household.' We are indeed moving towards a *self-service* economy, rather than a service economy.

Such fallacies persist because it's obviously very hard for most politicians to take on board the full impact of the new technology. If technological progress has always brought benefits in the past (and in their minds, it has), then why shouldn't it in the future? These irrepressible Micawbers assert that the present period of change represents no radical departure from the existing pattern; for them, it is just the next phase in a continuing, if not permanent, Industrial Revolution. Like garlic before a vampire, they flourish their malodorous promises of full employment before the onrush of the dreaded micro. Such empty words will avail them nought, for they are about to get chomped.

And it's not as if they haven't been warned. In *Sleepers, Wake! Technology and the Future of Work*, a swingeing indictment of these political relics, the Australian politician Barry Jones strongly argues the case that we must see this period as one of 'radical discontinuity', detailing the many elements involved in the adoption of new technologies that have no precedent in economic history. And a growing host of economists, trade unionists and social commentators in this country have outlined the revolutionary changes that the labour market will experience in the next decade. The consequences of this are daunting, for the effects of the recession will almost certainly be accelerated rather than ameliorated by the widespread adoption of microprocessor technology.

Let Them Eat Chips

The only serious argument against this view is that the new technologies will allow us to produce many new products, which in turn will mean more jobs. There may indeed be some potential in this respect, and it is certainly true that thousands of jobs will be created as the computer and associated 'software' industries expand. The UK Response to the *World Conservation Strategy* includes an important section on the future of work which emphasizes the potential of the so-called 'sunrise industries', such as energy conservation and bio-technology. But there is no chance that any of these developments could possibly compensate for all the jobs lost. Moreover, temporarily to boost employment through the manufacture and promotion of superfluous consumer durables, while thousands in this country still lack the basic necessities, and millions throughout the world still cannot get enough to eat, would merely confirm the moral bankruptcy of our embattled, inhumane industrial economy.

The erosion of job opportunities for the least skilled members of society, for the young and for the ethnic minorities reflects a longer-term change in the economy, the effects of which will become cumulative. Even the OECD has commented on some of the consequences of prolonged unemployment among young people – including the overall reduction of a society's skills and resources, and growing alienation of the young from the dominant social and political order: 'Everyone knows that youth employment is the biggest powder-keg in European politics. What we are all wondering is, how long is the safety fuse?'

But perhaps the most important consequence of all is something that still goes largely unnoticed. The unemployment problem is also a *distribution* problem. Since all industrial economies allocate wealth primarily on the basis of the job that a person has, mass unemployment aggravates existing inequalities of income to a quite unacceptable degree. As Hazel Henderson says, 'The microprocessor has finally repealed the labour theory of value; there is really no possibility of maintaining the fiction that human beings can be paid in terms of their labour. The link between jobs and income has been broken.'[2] One really needs to let that sink in, for it implies that only a fundamental restructuring of our economy, including the introduction of some form of national dividend or social wage, can offer any kind of promise that people will receive their fair share of the nation's wealth.

The challenge therefore is to achieve an orderly transition to an era in which we have to cope with substantially reduced opportunities for conventional employment. It is simply *not* an option at the moment to stand out against the introduction of the new technology. 'Develop or be damned' is all too literally true in this technologically besotted world of ours. Moreover, from an ecological point of view, there are considerable advantages to be derived from being able to generate increased levels of goods and services from significantly reduced inputs of energy and raw materials.

So what is the response of our politicians to all of this? Fantasies to the right, fantasies to the left, on into the valley of the dole ride the 1 billion. The fantasy of the right is that once government expenditure, taxation and inflation are sufficiently reduced, a born-again private sector will rapidly expand and so provide large numbers of new jobs. The fantasy of the left is that through massive government expenditure and rates of economic growth up to 5 or 6 per cent (I kid you not: it's there for all to see in Labour's *Alternative Economic Strategy*), millions of jobs will be created – and inflation can look after itself. The fantasy of the centre is based on an uncomplicated nostalgia for the good old days and the application of just enough grease to the rusty wheels of our economy to prevent it from collapsing for just a little bit longer. And when scorn is rightly poured on this fantastical hotch-potch, out comes the big one, the mega-fantasy of the 'leisure society'. 'We do not believe that work *per se* is necessary to human survival or self-esteem,' says no less an authority than Clive Jenkins. There will be more to say on this later.

Creating Inflation

There are many elegant theories to account for inflation and there are shreds of truth in most of them. I don't intend to join battle with any, but merely to point out that *whichever* theory you favour, you probably aren't getting the whole picture. Inflation is not so much an economic disease, to be cured by control of wages or of the money supply, as a symptom of much deeper problems. It may well be necessary to take a step back from the bewildering complexities of high finance and simply ask, why is life getting so much more expensive? For a start, everything seems to be so much more complicated, and it's simply more expensive coping with these complications. At the individual level, there are all sorts of things that superficially indicate a higher *standard* of living, but in reality merely increase the *cost* of living. If you're a commuter, or someone who has to install a burglar alarm, wear a

different suit for every day of the week, take tranquillizers, or
consult a lawyer, these things depress the real quality of life by
imposing on you additional financial and psychological burdens.

At the national level, the social costs of our present way of life
mount inexorably. The proportion of GNP that now has to be
spent on caring for the casualties (be they alcoholics, drug addicts,
glue sniffers, traffic accident victims, those suffering from stress or
psychological disorders), controlling crime, patching up the after-
effects of community disruption, protecting consumers and providing increasingly complex bureaucratic co-ordination has
reached an astonishing level. And though all these things are
counted in as positive contributions to our GNP (which is, don't
forget, taken to reveal levels of genuine prosperity), who can
possibly claim that they add to our real standard of living?

Industrial societies generate a bewildering array of these social
costs. The complexity of our way of life is so taken for granted
that many don't even realize the hidden burden it imposes on us.
Take an ordinary lettuce: it doesn't cost much to grow a lettuce,
just the cost of the seed and the tenancy of a little patch of land for
a short while. But for the average urban consumer it's not as easy
as that. First lettuces have to be picked, then wrapped in
individual cellophane wrappers, then crated up, loaded on to a
lorry, taken to the city, unloaded, processed through the books of
the wholesaler, picked up by the retailer, unloaded again, and
neatly arranged on the shelves of your local greengrocer. Is it any
wonder that even the humble lettuce comes a little expensive?

Then again, there are the costs of cleaning up the mess caused
by living the way we do. Not only the colossal amount of money
spent annually on the collection and disposal of rubbish, but also
the growing costs of pollution all have to be added to the
industrial bill. And the more you pollute, the more it costs. *All* of
these social and environmental costs are referred to by the
economists as 'externalities'. They don't really count, you see. Yet
they're *all* part of the true cost of production. Individuals, firms,
institutions, government agencies may all find ways of externaliz-
ing some of their costs by passing them on to each other, the
'system', the environment or future generations; but somewhere
along the line, someone pays. There are no free lunches on a finite
planet. Our *real* standard of living therefore depends on our

ability to keep down the costs of our social and industrial machinery.

Diminishing Returns

Conventional economists simply refuse to face up to the crucial relationship between the number of people living on the planet and the resources available for their use. When I was doing my economics A level, I can remember only too well that if demand for a commodity increased, or the supply shrank, then the price went up. I can't for the life of me understand why people don't realize that this simple truth remains at the heart of much of today's inflation. There's no dispute that demand *is* going up: not only are there more people, but there are more people demanding more as their standard of living improves. And as we've seen, there's no dispute that the supply of many commodities *is* shrinking as resources are depleted or become harder to obtain. It's hardly surprising therefore that evidence of rising *real* costs, and the greater effort required to maintain a given standard of living, is now apparent across the face of the planet. We are quite clearly experiencing diminishing marginal returns as the Earth's natural limits are approached.

And that's what it's all a question of: the Law of Diminishing Returns – or, for those who prefer a more accessible explanation, the accelerating slaughter of the geese that lay the golden eggs. As Lester Brown and the authors of many Worldwatch Papers have patiently gone on explaining, you can't just turn your back on this and hope the problem will go away. It won't; it will get a great deal worse. We've already considered the sort of pressure being placed on the biological systems that underpin our economies. In Worldwatch Paper 53, Lester Brown uses the example of the response of farmland to the application of fertilizer in attempts to raise productivity. During the 1950s, each additional million tons of fertilizer led to an increase of 11.5 million tons in the grain harvest; in the 1960s, it was down to 8.3 million tons, and in the 1970s to 5.8 million tons.[3] It's still going down. Quite soon, the cost of the fertilizer will be greater than the profit from the additional yield: the return will have diminished to zero.

Looking at this worsening population/resource ratio, we should be anticipating the day when 'global demand finally approaches the level not just of current supply, which has happened periodically throughout history, but of optimum feasible supply, which has never occurred before at the global level. The implication for prices is awesome.'[4] In the past, supplies could be increased to match demand, through fertilizers, irrigation schemes and new farming techniques; but in the future this will be much harder. Technological advances, leading to greater efficiency of resource use, can sometimes compensate for a decline in overall quality – but only up to a point. And the theory that we'll be able to substitute one ample resource for another diminishing resource is already a nonsense when simultaneous pressures are being felt across the whole range of global resources.

The same is true as regards oil and other sources of energy and raw materials. We now have to invest more and more capital in the process of getting our supplies of energy and raw materials from deposits that are either of lower quality or less accessible – with the obvious diminishing return. More and more energy is itself needed to extract the oil or the raw materials we require. Over the last fifteen years, the average pumping distance of one ton of oil in the USSR has increased from 650 to 2,000 kilometres.[5] Not only is off-shore drilling far more expensive anyway, but also, as we're now discovering in the North Sea, the use of secondary and tertiary recovery techniques (pumping in water, gas or air) raises costs enormously. We're up against the difference between *gross* energy and *net* energy. Gross energy consists of all those theoretical barrels of oil waiting to be guzzled up by us. Net energy is what we're actually left with once we've got it in a usable form; and given that it now takes the equivalent of millions of barrels of oil to get millions of barrels of oil, that is a fraction of the quantity theoretically available.

Eugene and Howard Odum, the American 'thermodynamicists', have calculated that the entire nuclear energy enterprise has yielded an absolutely neglible amount of *net* energy. So greatly is it subsidized by other sources of energy at every stage of the nuclear cycle (from the extraction of the uranium through to its reprocessing, from the construction of the reactor to its eventual decommissioning) that the net return from it has dwindled to the merest electrical pulse. It's worth a brief detour to contemplate the full implications of this.

Too Deep to Decipher

Engaging in any serious discussion on the costs of nuclear power is like conducting a three-legged race backwards through a mine-field. It is the fault not of the educated lay public that such should be the case, but of the CEGB's accounting practices, which amount to a deliberate campaign of misinformation. And don't imagine this is just another unfounded eco-groan. The recent reports of the House of Commons Select Committee on Energy and the Monopolies Commission take very much the same attitude, upbraiding the CEGB for misuse and abuse of statistics and criticizing its strategic thinking and seriously inaccurate forecasting.

Over the last couple of years there has been a significant shift in the whole tenor of the anti-nuclear campaign. The other worries remain, but the key concern now is *cost*. The reason for this is obvious: this Government's apparent irrationality in committing itself to ten 1,500 MW (megawatt) nuclear reactors before the end of the century at a time when we already have considerable surplus capacity. Such a programme will require staggering amounts of capital, perhaps as much as £30,000 million, a huge proportion of this country's total investment capability.

This is not the place to go into any detailed treatment of the costs of nuclear power. I can only refer you to the experts, particularly Walt Patterson's book *Nuclear Power* and Colin Sweet's excellent (and totally comprehensible!) pamphlet for the Anti-Nuclear Campaign (ANC), *The Costs of Nuclear Power*. Suffice it to say that whatever else you may have heard, it is without doubt a lie to claim that nuclear power is cheaper than all other sources of electricity. Little that the CEGB says can be taken at face value since its figures are manipulated in innumerable ways to conceal as much of the truth as possible. The figures often fail to include various hidden subsidies, or to allow for the effect of inflation on costs. They never include the costs of research and development, often leave out interest charges and exclude (or provide pathetically low notional figures for) the disposal of nuclear waste or the decommissioning of nuclear reactors. The CEGB's once proud, now pathetic claim that nuclear power would be 'too cheap to meter' has been replaced with a different slogan: 'too deep to decipher'.

No one disputes that nuclear reactors are much more expensive to build – even the CEGB concedes this! Consequently, the running costs have to remain much lower for nuclear power to remain competitive. The reality is that this has simply not happened: they've increased almost as rapidly as the costs of construction. Even the Magnox reactors (the first generation of reactors) have turned out to be more costly than coal-fired stations.[6] The second-generation Advanced Gas Reactors (AGR) have resulted in a disaster of quite historic proportions: the entire programme, which began in 1965, has accumulated a time overrun on construction of thirty-seven years! In Duncan Burn's submission to the House of Commons Select Committee, he estimated the real cost of the AGR programme at something between £8,700 million and £11,100 million: 'The only return on this investment – spread so far over fifteen years – has been the intermittent output of electricity from the two of the five stations which have worked, whose four reactors have averaged roughly 30 per cent of their design capacity since 1976.'[7] The two new AGRs at Torness and Heysham B will be the most expensive nuclear reactors anywhere in the world. And now (unless the Sizewell Enquiry comes up with the biggest surprise of all time!) we're going to get the Pressurized Water Reactor (PWR), a reactor with an appalling safety and operating record, that will cost even more than the AGRs to build. The spectre of Three Mile Island stalks the Suffolk coast.

It's worth remembering that none of this could be happening in the USA, where the nuclear industry faces a complete breakdown. There have been no new orders since 1978, scores of orders have been cancelled, and several half-completed reactors have been mothballed. By 1976, when the downturn started (*before* the accident at Three Mile Island), the energy utilities had independently seen enough to call a halt. Nuclear energy has priced itself out of the market, despite the dramatic rise in the price of oil. Yet while the Americans have learned the errors of their nuclear ways, we are still committed to nuclear power through a secretive energy monopoly and obsessive government pressure.

Scylla and Charybdis

The risks involved in nuclear power are considerable – so considerable that I personally don't want to see *any* nuclear reactors in this country at *any* price. But I am of course accustomed to the cost/benefit rationalizations that satisfy most people, and to them I say, given the risks, surely we should only proceed if the economic advantages are clearly demonstrable? If they are not (as would seem to be the case), why proceed? Because we *need* them, do I hear you say? The energy gap, and all that... This is *such* nonsense: there is *no* energy gap now (overall consumption today is less than in 1973), nor need there ever be. With an even halfway sensible approach to conservation and holding down costs, *demand need never rise again*. Moreover, despite artificial incentives, electricity is the least efficient and least competitive means of providing energy for most low-grade energy purposes. The all-electric economy is an absurd fantasy. At the moment, nuclear power produces 12 per cent of our electricity, which is just 3 per cent of our total energy use. Even with the Government's proposed expansion, this would only double to 6 per cent. The idea that nuclear power could provide a substitute for oil and gas is laughable.

So, not only are there grave risks involved, not only do we, the consumers, have to pay more because of it (as the only nation in Western Europe that's self-sufficient in energy, we still have the highest prices) – we don't even need it!

It's literally crazy. Yet it is the opponents of nuclear energy who are considered irrational, who are held to be beyond the pale, who are even considered by some to be subversives in the pay of the Kremlin! The most important conclusion to be drawn from this is to realize that the debate about 'need' is being conducted *outside* the realm of rational argument. The apologists of industrialism, intent upon their smog-bound horizons of continuing expansion, have impressed upon people that nuclear power is our lifeline to the future. In a mind-bending exercise of twisted logic, it therefore doesn't matter to the CEGB that it's more expensive; it doesn't even matter to Mrs Thatcher, which might otherwise be considered strange in one so hostile to all other monopolies and wasteful extensions of public expenditure. The decisions are therefore not made according to economic criteria, but as a result of institutional and ideological pressure.

This is but one example where the apparent rationality of our decision-making has completely broken down. In many other areas of economic policy, the traditional guidelines and theoretical explanations have become quite redundant. It used, for instance, to be held that unemployment and inflation traded off against each other: that as one went up, the other went down. We now have structural unemployment (which is clearly not just a temporary aberration) and structural inflation (which will clearly not be 'cured' by any amount of financial wizardry) existing side by side in all industrial economies. It used to be held that economic growth was the only cure for unemployment. We now see that unemployment can actually be *caused* by economic growth. Conventional economics has all but collapsed, its theoretical basis now bankrupt, its practice quite unable to handle the unfamiliar problems that the world now faces.

As regards inflation, for instance, all the pressures are now lining up in such a way as to confound conventional approaches: resource limitations, diminishing returns, social costs, institutional inertia, human frailty and artificial expectations. We live in an *inherently* inflationary society quite simply because our kind of prosperity can be achieved only at the expense of the natural capital of the planet. The implications for this country are grave. Though temporarily self-sufficient in energy supplies, we remain dependent on the rest of the world for much of our food and raw materials. The danger is that with overseas markets hardening against us, our manufacturing and service industries will fail to earn enough foreign exchange to maintain our primary base through imports.

By disregarding the biological and physical systems that underlie all economic activity, we have thrust ourselves between a contemporary Scylla and Charybdis, the rock of mass unemployment and the whirlpool of permanent inflation. And unlike

Ulysses, the odds are that we shall be incapable of finding a way through. The sails can be trimmed; the suspicion is that the ship cannot be steered. For the driving force that powers our frail vessel, the ideology of industrialism, has taken control both of the direction and of the speed at which we travel. In so doing, it has induced so far-reaching a sense of alienation that most people do not even understand our predicament and, even if they did, would probably feel there was nothing they could do about it anyway. The iron embrace of industrialism has paralysed our minds and corrupted our souls.

7
Alienation is a Way of Life

'Sorrowful Drudgery'

Ever since Marx first started using the word in the nineteenth century, 'alienation' has become a popular critical term. The number of contexts in which you now find it has increased enormously, and though some may say this is just the consequence of typically sloppy usage of the English language, it seems more likely that its current familiarity acccurately reflects the extent to which alienation has become a part of our lives. I shall be using the word to indicate that sense of estrangement people experience between themselves and their work, their own health, their environment, and the workings of their democracy.

Much of the responsibility for this may be laid at the door of a work ethic that no longer works. Torn and tattered though they now are, our attitudes to work go back a long way. The God of the Old Testament imposed work on Adam and Eve as a punishment for eating the fruit from the Tree of Knowledge. There's an interesting similarity between this and the myth of Pandora, the first woman created by Zeus, who was given a box and told on no account to open it as it contained all the ills with which the human race might one day be afflicted. Well, she did open it – and all the evils flew out, including work, the Greek word for which is *ponos*, from which we derive such words as pain and punishment.

I don't suppose that old ogre Calvin knew much about Pandora, but he certainly regarded work as a 'punishment', something that must be suffered for the good of one's soul, an experience of self-denial rather than self-discovery. We are the heirs to that arid tradition, now subtly blended with the utilitarian values of the Industrial Revolution. We have made of work a unique kind of 'sorrowful drudgery', as William Blake referred to it. And we've managed to do so in the most extraordinarily irrational kind of way.

On the one hand, we look upon it as a cost or a burden which allows one to achieve certain benefits and pleasures. It's a necessary evil, nothing more than a way of putting money into people's pockets so that they can operate as effective consumers. The nature of the work done, the actual activity itself, is irrelevant; it is, after all, merely a means to an end. Yet on the other hand, it still remains the case that people are evaluated by what they *do* rather than what they *are*. Our self-esteem and the opinion of others is often determined by our job. The extreme position here is that any work, however degraded it may be, is good, while being without work is bad.

The consequences of this ambivalence for old people, for those who stay at home and for the unemployed are considerable. A disproportionate number of people die shortly after retirement, as if they felt themselves to have become obsolete. Those whose work is in the home are often looked down on, as if their life lacked any real sense of meaning or purpose. For the unemployed, this can amount to a form of 'social death'. They are forced by the compulsory work ethic to seek jobs that no longer exist while blaming themselves that it is somehow their own fault. Nor can they enjoy the compulsory leisure which is forced upon them, for in today's world it seems that employment and leisure are in a state of mutual dependency: you need one to enjoy the other.

We are today beginning to hear more about 'socially responsible production', but we still don't hear enough about the vast amount of 'socially *irresponsible* production' that characterizes our economy, degrading hundreds of thousands of workers both mentally and physically, while at the same time contributing to the destruction of our environment and life-support systems. The 'anaesthetic' of a well-filled wage packet fails to compensate for the damage inflicted through fragmented, dehumanized, alienating work. As Roszak points out, it's not possible to redeem such work by enriching it, restructuring it, socializing it, nationalizing it, municipalizing it, decentralizing it or democratizing it. It remains bad work. There's something intensely pathetic about that oft-heard cry, 'What do we want? JOBS! When do we want 'em? NOW!' Politicians and chanters alike are living in a distorted dream world, not only because the age of full employment is now of greater interest to historians than to politicians, but also because there's nothing enlightened, radical or responsible about the obstinate promotion of archaic notions of 'full employment', while paying so little regard to the *quality* of the work involved.

Our work ethic has become the Achilles heel of industrialism: either we state that it remains an important and useful attribute of society, in which case we must continue to create jobs regardless of the cost to people or planet; or we reject it, in which case we must find an alternative. It seems to me essential that we must indeed ditch the old work ethic, for reasons that I've made clear; but since work is a fundamental part of the process of self-definition, we must invent a new work ethic – or rather, rediscover a much older one. And this is exactly the political challenge that the established parties refuse to take on. Their history and all their experience is rooted in the development of an industrial way of life; the alienation that has resulted from that is one of *their* making.

Alienation at Work

The Industrial Revolution changed the whole nature of work. The division of labour, that unique creation of the industrial process, totally supplanted the sort of craftwork in which the individual was responsible for the whole production process and not just a part of it. This has been hailed as the key to economic progress, yet it is in fact the cause of much of the alienation of modern life. The specialization that resulted from it was achieved only at the cost of a general 'deskilling' of people; creativity and inventiveness became utterly remote from the process of work, allowing the individual to become fragmented along with the whole complex economic system.

As we now know, the division of labour was the beginning of

the end for the unskilled and semi-skilled worker, since the highly repetitive tasks created as a result of it can so easily be automated. It's worth remembering that from the very onset of the Industrial Revolution, machines were introduced not as a way of enhancing skills, but as a means of displacing people. There was little concern for the dignity and integrity of people's labour – the need for 'right livelihood' was considered totally irrelevant to the pursuit of greater productivity and profit. Work itself was subordinated to these 'higher goals', and generations of new machines were promptly incorporated into an already distorted work ethic. Since to the employer the ideal was maximum output without employees, labour simply became an item of cost to be reduced to the minimum; to the employee, the ideal became to achieve the maximum income for as little work as possible.

The same tendencies are at work today. Computers and microchip developments are specifically intended to displace labour, still in the name of higher productivity and greater efficiency. Instead of the new technology serving as a force for liberation, for which it undoubtedly has the potential, there is a terrible danger that it will merely increase the degree of servitude by which people allow themselves to be bound. The push for labour-saving technology comes from a very powerful minority, who are able, through the selective funding of scientific research and the even more selective implementation of technology, to channel that technology in ways which benefit them rather than society as a whole.

In his book *Architect or Bee*, Mike Cooley has drawn our attention to some of the dangers of the new technology: the increased stress from being subjected to work which is fragmented or demands an ever greater tempo; the historical tendency gradually to 'de-skill' all work; the destruction of individual creativity that this implies, and the loss of the means by which we build up the next generation of skills. He goes on to make it plain that 'control' has been as great a stimulus to technological change as the pursuit of greater productivity, warning us against the illusion that science and technology operate as neutral forces in society, and demonstrating time after time the gap between that which technology could provide (its potential) and that which is actually does provide (the reality of our industrial society today).

The fact remains that the vast majority of people have still not understood the true nature of alienation caused by industrialism. It is not alienation from the *means* of production or even from the *fruits* of production that really matters, but alienation from the *process* of production. The left has simply got hooked on the wrong thing. The 'socialization of the means of production' is all but irrelevant if the process remains unchanged. This alienation, characteristic of all industrial systems, capitalist or communist, is the key to understanding the kind of changes we are going to have to make.

That is why we have to be very cautious of certain proposals for industrial democracy or work enrichment. The next phase of industrialism will almost certainly take on board many such notions, for it can be demonstrated that such 'reforms' more than pay for themselves in terms of productivity. But any such improvements in working conditions or participation, though welcome in themselves, will still be constrained by the same imperatives of industrialism. They will not address the root cause of the problem: the alienation brought about by rigid hierarchical structures and the lack of opportunities for genuine participation.

It is also the reason why we have to be so cautious of those who push the 'self-evident merits of a post-industrial leisure society'. It has become a powerful image in popular mythology that progress lies in the relinquishing of *all* work to machines. I find such visions deeply disturbing. In his essay, 'Economic Possibilities for our Grandchildren', John Maynard Keynes referred to 'technological unemployment' as a 'temporary phase of maladjustment. All this means in the long run is that mankind is solving its economic problem...' In the narrow sense of industrial efficiency, the rationalization of production and the maximization of profit, this may very well be true. In every other respect, such a 'solution' would literally tear society in two, with a minority working in very well-paid capital-intensive jobs, and the majority forced to become ever more dependent on the state. There is no question that such a tendency would accelerate the development

of a highly centralized, authoritarian corporate state. Alienation
would be institutionalized as the dominant feature of society.

Cure Crazy

It's one of the more peculiar by-products of twentieth-century
industrialism that we have become so alienated from our own
bodies. The World Health Organization rather ambitiously
defines health as 'a state of complete physical, mental and social
well-being, and not merely the absence of disease or infirmity'.
And its absolutely right, though there's mighty little evidence
that such insight has percolated down to politicians and health
officials today. In this country, government officials actually take
pride in the fact that *more* patients are now being admitted than
ever before, and that *more* outpatient and emergency treatments
are now being provided than a few years ago. It would seem that
the less healthy we are, so much the greater is the success that may
be claimed on behalf of the National Health Service (NHS). The
ultimate goal, presumably, is to make patients of us all.

Stated in such terms, the crisis in the NHS is literally intract-
able. We shall always be spending more; and we shall always need
to be spending yet more again. For all the 'progress' in modern
medicine, for all the new drugs and impressive new technology,
for all the astonishing insights into the 'mechanics' of the human
body, the growing gulf between cost and effectiveness has
precipitated a major crisis in health care throughout the
developed world.

Yet it is abundantly clear that our health is predominantly
determined not by medical intervention, but by our environment
and our behaviour. The causes of ill-health are not extraneous to
society; the sharp decline in infectious diseases in the nineteenth
century was due far more to improvements in nutrition, and in
hygiene and sanitation, than to the rise of modern scientific
medicine. Today, in a socio-economic system that is fundamentally pathogenic, no one should be surprised by increases in the
chronic and degenerative 'diseases of civilization' – a 'civilization'
in which millions are the victims of stress, pollution and drug
abuse, are overfed and under-exercised, are addicted to cigarettes,

alcohol and convenience food, and are expected to thrive on mindless work and passive leisure.

In all Western countries, just two disease categories account for more than two-thirds of all deaths: cancer and diseases of the heart and blood vessels. Many of these deaths are premature and not just people dying naturally of old age. Coronaries and strokes account for half the deaths of people in Europe and North America. Interestingly enough, this proportion has not increased at all over the last fifteen years or so, and there's little doubt that awareness of the different lifestyle factors involved in heart diseases has enabled people to do something for themselves.

Unfortunately one cannot say the same about cancer. Looking at the incidence of different types of cancer among different societies, it is clear that environmental factors, particularly diet, play a huge part in this disease. In 1977, the US National Cancer Institute (NCI) estimated that as many as 60 per cent of all cancers in women and 40 per cent in men were significantly influenced by dietary considerations. It seems then utterly extraordinary that so little attention is paid to this aspect of the disease. The NCI itself, with an annual budget of well over $1 billion, spends less than 1 per cent on research into nutrition. The sad reality is that cancer scientists, in line with their colleagues throughout the medical world, are 'cure-crazy': they work on the assumption that cancer is due to a single cause which can be dealt with through a special cure. Their careers and their prestige are all wrapped up in such an approach. Despite the fact that a large proportion of premature cancer deaths could be prevented, prevention is low on the list of priorities of the present cancer establishment.

The fact that we seek so desperately for illusory cures is attributable to the narrow conceptual basis of modern medicine, or, as George Engel writes, to 'the notion of the body as a machine, of disease as the consequence of breakdown of the machine, and of the doctor's task as repair of the machine'. The increasing dependence on specialized and high-technology medicine has encouraged doctors to deal with particular parts of the body, often forgetting to deal with the patient as a whole. By reducing health to a series of discrete mechanical functions, modern medicine seems concerned more with *illness* than with the *person* who is ill.

Despite the obvious contradictions and absurdities built into this approach, one hears few voices raised in dissent. The NHS is now structured in a fragmentary, over-centralized and paternalistic way. The notion of individuals taking responsibility for their own health has been almost entirely lost. What doctors implicitly condone the advertisers explicitly exploit, sustaining the idea that however unhealthily we choose to live, there's always an instant, glossily packaged remedy available. The Health Education Council's budget is trivial compared with the constant and ever more sophisticated promotion of ill-health in our industrialized society. Even the most ardent practitioners of self-help health care find it harder and harder to avoid environmental pollution or the adulteration of their diet through the presence of potentially damaging chemicals. To spend billions every year on increasingly forlorn attempts to patch up the victims of our industrial society, or to cure disease and illness that could largely have been prevented in the first place, can hardly be described as a 'health' policy.

Crumbling Communities

It's in the inner-city areas that the most debilitating effects of bad health and unemployment have really taken their toll. For it is here that the forces of social disintegration line up with unemployment to make life such a misery for so many. Many of our cities are sorry places today; despite large sums of money poured into the central areas of big cities, the problems have merely got worse. The causes are all too obvious, stemming from an erosion of traditional community and family responsibility, the exodus of industry, and totally inept social planning.

The massive social experiment of urban resettlement has proved a disastrous failure – and many new development programmes continue even now to add to the burden. The social bonds that foster genuine cohesion and mutual support are torn apart. Everything and everybody is 'zoned' in an attempt to organize everything rationally: live here, shop there, play somewhere else and work wherever you can get it. Many traditional sources of employment are thoughtlessly destroyed;

corner shops, small businesses, workshops – everything has to go as part of the great plan. As the mainstay of the local economy collapses, more people move out, creating ghettos for the disadvantaged, the old, and the unemployed.

You can't just dissect communities like this; such a fragmenting of social relations goes completely against the grain of life. People become 'decultured', and as the level of alienation rises, so too does the likelihood of violent confrontation. It's bad enough having to live in a city anyway, with one's individuality crushed among the mass transport systems, the mass culture, and the mass processing of human beings as statistical abstractions. But when the city breaks down, then the writing is truly on the wall.

They're doing their best, of course, those planners and architects and civil servants. Within their own narrow framework of understanding, they can justify what they do as a totally rational response to a set of complex problems. But somewhere along the line they seem to forget that they are dealing with people and with all that chaotic cultural rag-bag of people's values, their sense of history, their eccentricities, their relationships with others, and above all, their sense of place. These things – the human ecology of a city – simply don't register in the contemporary planner's rationale. And in the meantime, if they can, people move out into the suburbs and join the hectic helter-skelter of sorely afflicted commuters. Because city life has broken down, people flee the city – just as people flee the land now that rural life has broken down.

Here again, the planners have a lot to answer for. (I don't really mean to victimize individual planners, but rather the lunatic system within which they are forced to operate.) Over the last twenty years our rural communities have gradually wasted away; farm workers have become an endangered species, comprising a mere $2\frac{1}{2}$ per cent of our present workforce; small farmers have been displaced as the move towards larger and larger units continues unchecked, despite clear evidence that this does nothing to enhance real productivity. Transport services are 'slimmed down'; village schools and post offices are 'reluctantly closed'; farm buildings and cottages remain empty: 'We have no choice,' say the planners.

But they do. For a start, they should get rid of this absurd nonsense of applying uniform policies to all parts of the country, regardless of local circumstances, traditions and experience. Secondly, they must understand that the only way to repopulate rural areas is to stop treating them like the poor cousins of our urban industrial society: just what kind of idiocy is 'rural industrialization' anyway? The problem is that they're preoccupied with creating *jobs* instead of providing opportunities for work and self-employment. They also have a totally inappropriate 'hands-off' attitude to the land. Of course it's important to limit urban sprawl, but this does not mean that the countryside should be set aside exclusively for agricultural uses. For centuries the countryside has bustled with people involved in every conceivable kind of activity, some linked to agriculture, but many not. Planning refusals are often made on the grounds of 'unwarranted intrusion into areas of unspoiled countryside'. But as many have pointed out, unspoiled countryside means more than vast depopulated farms, second homes and coachloads of tourists. Without people living and working there in every natural way that you may care to think of, the countryside is dead.

In the last century, Disraeli continually warned his party that if the 'outlook of the counting house' were applied to rural matters, then disaster would be the inevitable result. An important part of his philosophy of 'One Nation' was a healthy balance between the rural and the urban. But to achieve this our attitude to the land itself must be radically changed. Only a tiny proportion of people in this country actually own farming land, and they have become a very privileged elite. It's clear to ecologists that the land cannot be owned in the sense that one owns a car or a dishwasher. The land is part of our *common wealth*; though we may be free to acquire the products of our own making, we are not free to acquire the Earth in this way. The present system denies people their natural birthright of access to the land, and is incompatible with the ecological emphasis on stewardship rather than ownership, on recognizing the land as a common heritage to be cared for on behalf of the community and future generations. The monopoly control of land ownership must be brought to an end, not through the fossilized notion of nationalization, which might well make things worse rather than better, but through measures of radical *communal* reform.

Inhuman Scale

Ever since the publication of E. F. Schumacher's *Small is Beautiful*, concern about the scale of things is the one item of green thinking that seems to have sunk more deeply into the troubled psyche of people than almost any other. The tendency of industrial cultures to extend the 'economies of scale' into every area of life has become a matter of concern even to ardent industrialists. But such concern is often superficial, for it remains a logical and *necessary* consequence of an industrial ideology that all forms of human activity should be encouraged to grow and expand. The ideology of industrialism depends upon the interaction of masses – mass production, mass consumption, and masses of people. As a result, all urban-industrial economies are beginning to 'coagulate into a single, planet-wide society', as Roszak puts it, dominated by 'insensitive colossalism'. An essential element of the green critique of industrialism is that the very bigness of things is one of the main reasons why we not only fail to meet the needs of most people, but end up destroying the planet even as we fail.

A word of caution: as we shall see later in chapter 12, 'small is beautiful' is a very limited slogan – and it has been misused by a lot of very naive people. It's not simply a matter of size in quantitative terms, it's as much to do with the *quality* of scale. Whatever size it is that takes away our dignity, makes us passive recipients rather than active participants, makes us dependent rather than self-reliant, alienates us from the work we do and the people we live with – *that* is too big. To try to determine an appropriate size for each and every human activity is like trying to regulate the number of angels that may sit on any one cloud.

Notwithstanding such caution, it's becoming pretty obvious that at the same time as we approach various environmental and biological constraints on growth, so we are reaching certain institutional limits imposed by the growing incompetence and declining performance of our bureaucracies. The levels of inter-dependence and complexity are now so great in many bureaucracies that even the ablest of decision-makers within them are quite overwhelmed. The costs of co-ordinating this complexity are considerable. The larger an organization or bureaucracy becomes, the more rigid and inflexible is it, and so much the less scope is there for creativity and divergent thinking. Similarly, the larger it becomes, the more likely is it that standardized, depersonalized methods of operation will increase the amount of alienation people feel.

In short, we're up against yet another application of the Law of Diminishing Returns: at a certain optimum size or scale, no further advantages can be derived from further increases in size. Thereafter, *diseconomies* of scale will rapidly become apparent. That, one might think, is fairly obvious, and yet most bureaucracies are quite incapable of any self-monitoring discipline in terms of recognizing what their optimum size may be. As the excellent radio and TV series *Yes, Minister* has so humorously exposed, the bureaucratic imperative is one that brooks no compromise! Size itself is a source of power, status and prestige; to think of thinning oneself down in such a context runs counter to the whole trend of government and business practice.

These days bureaucracies are up against particular problems as regards the handling of information. In the rhetoric of slick self-advertisement, we live in an 'information-rich, dynamic age'; that may be so, but we remain embarrassingly 'knowledge-poor'. Just because the computers are churning out great wodges of information, that does not mean to say that knowledge is growing proportionately. This is of particular relevance to the workings of our democracy. With fewer voters actually able to understand what's going on in so complex and 'sophisticated' a society, and with Parliament increasingly in the hands of experts and specialists who look to Parliament merely to rubber-stamp the product of their expertise, the very nature of our democratic decision-making process is becoming sadly devalued.

It's already devalued enough. The sense of alienation that puts a distance between the people and the democracy that serves them has deepened considerably. Despite, or perhaps because of, our great technological power, these are times of great powerlessness for the majority of people. Our 'caring society' has become so remote and impersonal that those at the bottom of the heap have come to feel that the system now cares nothing for them.

In Defence of Democracy

A profound dissatisfaction has sprung up as regards the distinction between *empty* participation and *genuine* participation. Among examples of the former are to be numbered almost all show-piece enquiries (such as Windscale and Sizewell), much of the so-called 'consultation' involved in planning procedures and other local government activity, and many aspects of representative democracy. As Stephen Cotgrove says, 'Where belief in the reasonableness of the political system, and its openness to reasoned argument and debate, breaks down, the normal channels of petition, protest and pressure group tactics come to be seen as inadequate.'[1] If the main purpose of democracy is to ensure that Government, Parliament and its various bodies, corporations and committees are as answerable as possible to the people, so that individual citizens may have the maximum say in government, then democracy in this country is failing in a number of crucial respects.

One can't help but reiterate that the most grievous of these is the lack of proportionality between the number of seats a party ends up with in Parliament and the votes cast for it. No post-war Government has had the support of even a simple majority of votes cast at the general election which saw it elected, yet despite that, successive Governments have attempted to make sweeping changes in the nature of society and the direction of the economy. The present Government has 61 per cent of the seats in Parliament from 44 per cent of votes cast and just 32 per cent of those eligible to vote. (The media all referred to this as a swing to the Conservatives, yet their vote actually went down by 1.6 per cent.) I'm very conscious of the element of 'special pleading' involved here; it's true that the Green Party is unlikely to make significant progress until some form of proportional representation (PR) is introduced into this country. None the less, I can't help but suppose that whatever my political persuasion, my concern about the obvious unfairness of our present system would remain – even though my indignation might be more restrained. For the key issue at stake is that the existing system is doing us *all* a grave disservice; the loss of credibility incurred through such ridiculous results exacerbates the long-term threat to the resilience of our democracy.

To us it seems just as important that PR should be introduced at the *local* level, a reform which we see as an essential part of the struggle to redress the balance between local authorities and central government. For all their fine words about decentralization, central government, be it right or left, has increasingly overturned the powers of local democracy where it has suited it. Controls over rate levels and spending levels are becoming so rigid that we are beginning to see a system of local *administration* taking shape rather than local *government*. There is no recognition of regional (or indeed, national) diversity within the UK, and since the fiasco over the setting up of Scottish and Welsh Assemblies back in 1979, little has been heard about the need for devolution of a wide range of fiscal, legislative and administrative powers – little, that is, in the south of England.

There's a totally unwarranted complacency in this country about the health of our democracy. The latest symptom of this was the decision to raise the deposit for parliamentary elections to £500 from its former level of £150, on the ground that too many candidates were using the electoral process for cheap publicity. This is a most dangerous and retrogressive development, for it means that one's suitability as a candidate is to be determined by the amount of cash one has. Smaller and less wealthy parties will be forced out of the electoral process, as too will many serious independent candidates.

If the aim is to discourage frivolous candidates, then why not simply raise the number of signatures required on the candidates' nomination forms from 10 to 100 or even more? This would be a far more useful proof of 'fitness to stand', as it would demonstrate beyond reasonable doubt the seriousness of the candidate concerned. Yes, it might take a bit longer checking all the signatures, but that's infinitely preferable to conducting an electoral means test. It is vital that there should be clear and easy access for all prospective candidates to the electoral process; to arrange otherwise will be to strengthen the hand of those who dismiss the workings of democracy as a sham.

It seems peculiar that this country of all countries should just take on trust the essential freedoms and rights on which a democracy depends. It is well known that the police and security agencies are already heavily involved in computer and electronic surveillance and that phone-tapping is carried out on a considerable scale. As our industrial system becomes more dangerous, more complex, more alienating, so the measures taken to retain control over it are likely to become more authoritarian. It is already clear that the dual dependence on nuclear weapons and nuclear power has upset the balance between state authority and the rights of the individual. A 'plutonium culture' allows for no other choice; in 1976, a force of special constables under the direct control of the atomic authorities was set up to guard nuclear facilities. They have also been mandated to check up on 'dissidents', those who seek to stop the further development of nuclear power. The point is that plutonium can *only* be protected by police-state methods; intelligence-gathering among the civilian population, and other encroachments of civil liberties, are seen as necessary and rational measures within such a framework. The Flowers Report made it absolutely clear that to develop the plutonium economy would make 'inevitable' the erosion of basic freedoms.

In order to protect democracy, one must simultaneously protect the rights of individual citizens. That requires two vital reforms: first, a comprehensive Data Protection Bill, which must go a great deal further than the present Government's pathetic attempt at minimum legislation. Secondly, we must ensure genuine freedom of information in this country, so that Ministers and whole departments are no longer able to hide behind the Official Secrets Act. This isn't just a question of the occasional scandal caused by section 2 of the Act; the 1984 Campaign for Freedom of Information has revealed the extent to which secrecy permeates *every* facet of our industrial society. For instance, a recent report by Friends of the Earth for the Campaign demonstrates that the path to adequate environmental protection in this country is blocked by the unacceptable levels of secrecy demanded by industry and supported by government agencies.[2] Behind these barriers there lurks a web of corruption, incompetence and disregard for public safety and individual rights – further compelling proof of the breakdown between ordinary people and the system that claims to defend and represent us.

8

A System without a Soul

Alienation has indeed become a way of life, feeding on the ethical and spiritual vacuum at the heart of our society. Industrialism is the ruling ideology that makes the wheels of politics turn, but that alone cannot account for the state of global disorder that confronts us now. To my mind, this is where Fritz Schumacher was a *real* radical, for his critique of industrialism was rooted in his concern at the loss of a spiritual dimension in most people's lives. Without some sort of spiritual underpinning, he saw little hope for those who are daily bludgeoned with economic necessities and seduced by the delights of materialism.

There is no more important expression of this than in his essay 'Peace and Permanence'. In this he overturns what little is left of the average Western Government's sense of vision or ultimate purpose. For it is still almost universally held that the surest foundation for peace would be universal prosperity: the road to peace is the road to plenty. Schumacher writes: 'I suggest that the foundations of peace cannot be laid by universal prosperity, in the modern sense, because such prosperity, if attainable at all, is attainable only by cultivating such drives of human nature as greed and envy, which destroy intelligence, happiness, serenity and thereby the peacefulness of man.'[1]

This simple but to me utterly compelling insight has a long and powerful tradition in the works of many political and spiritual authors. The particular trigger in Schumacher's case was the article by J. M. Keynes to which I've already referred, 'Economic Possibilities for our Grandchildren'. In this, Keynes looked forward to a time when all our economic problems would be solved and we could:

> once more value ends above means and prefer the good to the useful. But beware! The time for all this is not yet. For at least another hundred years we must pretend to ourselves and to everyone that fair is foul and foul is fair; for foul is useful and fair is not. Avarice and usury and precaution must be our gods for a little longer still. For only they can lead us out of the tunnel of economic necessity into daylight.'[2]

These are chilling words. They remind me of those patronizing dismissals from people with whom I often share public platforms, who, having been obliged to acknowledge the seriousness of the ecological crisis we face, still assert that it is a problem for the future rather than for now. If economic progress in the future is attainable only through cultivating the powerful human drives of 'greed and envy', then the road to Armageddon will indeed be paved with gold. For a few.

As if *All* People Mattered

Perhaps the most painful experience of my eight years in the Green Party was acting as agent in the Croydon North-West by-election in 1981, just at the time when the Alliance band-wagon was spluttering into fitful life. Spurred on by the success of the Radical Party in Italy, which had actually managed to get the European Parliament to take Third World issues more seriously, we decided to base our campaign on the links between world poverty and arms spending. Our leaflet was very different from the usual run-of-the-mill election leaflet; our canvassing was fairly thorough; our press coverage was no worse than usual; and to lend actions to words, the whole 'campaign team' lived on a minimal Third World diet for the duration of the election. We got precisely 155 votes.

It may all sound very naive; in retrospect, it probably was. Who were *we*, after all, to ask the voters of Croydon to lift their eyes a little from the banalities of British politics? The derisory

vote I could have lived with – I'm used to it by now – but it was
the levels of ignorance and sheer indifference that appalled me.
Canvassing became a masochistic confirmation of everything I'd
always been told: that politics amounts to no more than answer-
ing the question 'What's in it for me?' As far as the voters of
Croydon were concerned (and I don't imagine they're any better
or worse than any other voters), the Third World did not exist.

It was a hard lesson, and one that I still haven't learnt to live
with. Pragmatists though we may sometimes be, it's just not
possible for greens to turn their backs on the problems of the
Third World. We just don't understand how any politician's
'vision' of a desirable future can countenance such utter degradation for so many. It's all been documented a thousand times: how
the one-quarter of the world's people that lives in the countries of
the North (excluding China) consumes 80 per cent of total global
resources, leaving the three-quarters that live in the South to share
out the remaining 20 per cent; how 800 million people live in
'absolute poverty', defined as 'a condition of life so characterized
by malnutrition, illiteracy and disease as to be beneath any
reasonable definition of human decency'. The basic problem is
simple: while the unbridled affluence of a minority of humankind
is for ever on the up and up, the inexorable impoverishment of
the majority is somehow taken for granted. Not only is that
utterly and insupportably unjust, it is also ecologically unsustainable, and carries within it the seeds of apocalyptic disaster. There
is something totally irrational about a system which succeeds in
satisfing the wants of a minority only by denying millions even
the chance to survive. Such a system might well be said to have
lost its 'basic legitimation'.

The conventional, 'caring' case for action is based on the notion
that development in the Third World depends on rapid economic
growth in the North, so as to provide bigger export markets for
Third World countries and to ensure that a higher proportion of
GNP can be spent on aid. To us, this approach is almost as bad as
doing nothing. It's amazing how the myth still lingers on that
through extensive investment in Third World countries, we are
actually helping them to develop. The reality is that in such
circumstances, those countries will be forced further and further
into *under*development.

Global Rip-Off

Year by year the figures reveal that the developed world takes far more out of the Third World than it puts in, largely in terms of repatriated profits on foreign investment. It is impossible for us to maintain our levels of affluence without a massive appropriation of Third World resources, as was always the case throughout the period of colonialism. On its own, such a process might still have permitted many Third World countries to make modest improvements in their standard of living; they could never have caught up, of course, but the gap might have narrowed. As it happens, even this consolation is not available for us to take comfort in: even with significant improvements in GNP and rates of growth, the *kind* of development we are promoting has proved disastrous. To quote a World Bank report: 'Although there has been encouraging economic growth over the past three decades, a very large portion of their people have not shared in its benefits. On average, the poorest 40 per cent of their societies is not much better off than it was.' And that means barely surviving: a billion people live in countries where the average income per annum is $220 or less. Even with real growth at 2 per cent, that means an increase in per capita income of around $5 a year. Is that really what we mean by development?

Yet again, the vastly overworked theory of 'trickle-down' is seen to be quite fraudulent. Far from alleviating the worst manifestations of poverty, the sort of foreign investment generated by contemporary industrialism merely aggravates them. If anywhere, the wealth of the Third World trickles down to *us* rather than to its own people, though most of it goes to building up privileged elites within each Third World country and swelling the vast profits of the multinational corporations.

It is the power of those elites that causes many of the problems. They must surely be able to see that their countries are being ripped off left, right and centre, and yet they choose to do nothing about it. Most international economic theory is still based on our experiences during the early days of the Industrial Revolution, and is now quite inappropriate. The notion of 'comparative advantage', for instance, with each country producing what it's best at and exchanging it for what it needs in the world market, may once have led to huge increases in world production, but is now little more than an excuse for systematic exploitation.

It's a dual process. First, Third World countries generate most of their exports through primary commodities, particularly food and raw materials. Over the last twenty years, the number of food crops grown exclusively for export has risen steeply; dubbed 'strawberry imperialism', this sort of development means we get our luxury fruit and vegetables all the year round, while many of the countries most involved in such exports have become dependent on *imports* of grain and other basic foods merely to survive. We have turned a huge acreage of prime Third World farming land into our very own out-of-season vegetable plot. In its desperate need for foreign exchange, it now has to export more and more to get the same amount of goods in return. There are wide fluctuations in price from year to year, often caused by speculation on the commodity market, and this further destabilizes its already vulnerable economies.

At the other end of the process, Third World countries are seen by Western exporters as a huge, expanding market; one-third of exports from industrial nations go to the Third World. Many of these products are useful, but many are not. Some are downright harmful, such as refined foods, drugs, powdered milk, cigarettes and pesticides – let alone the huge volume of arms of every description. Pressure is often exerted on Third World countries to open up their markets (through conditions on loans, for instance), even as we ensure that ours remain closed to their 'cheap imports' through discriminatory tariffs and quotas. All kinds of incentives are used to 'hype' the sales of dubious products and to create totally artificial needs, often with the blessing of the leaders of the countries concerned.

The key to this pattern of development is the involvement of the multinationals. The 'multinationalization' of the world economy continues unchecked, and they now control between one-quarter and one-third of total world production. They are keen to exploit the cheap labour of the Third World, and to benefit from their 'better labour discipline', minimal safety regulations and lax pollution controls. They are keen to facilitate the extraction of Third World countries' raw materials. They are keen to build up powerful middle-class cliques in the familiar

urban-industrial mould. The investments they make are therefore
not the ones which would be made if the primary concern were
the well-being of Third World people.

That is the main reason why ecologists are firmly opposed to
the initiatives of the Brandt Commission, *North–South* and
Common Crisis. Both reports are obsessively dependent on the old
model of expansionist industrialism; they are primarily concerned
with the preservation of the existing world order, to which the
continuation of world poverty has become a serious threat. The
poor should be helped because the system can't function without
them.

But in the meantime, the cycles of poverty, exploitation,
illiteracy and disease go on. The poor have little time or
inclination to worry about global environmental trends, and yet
in many ways they are more affected by the ecological crisis than
the affluent who can just drive away from it. Many Third World
people are forced by circumstance to destroy the very resources
on which they depend. The metaphor of eating the seed corn for
next year's planting is duplicated time after time throughout the
Third World. There can be no clearer demonstration that those
who are working for a better environment must simultaneously
devote themselves to working for social justice. There is not only
the moral imperative that compels us to seek ways of sharing the
world's weath more effectively; there is the ecological imperative
to remind us that the protection of the Earth's natural systems is
something we *all* depend on.

All Creatures Great and Small

As we have seen, the degree of that dependency is something that
people are only just beginning to take account of. It is extraordinary that we should be engaged in so devastating an alteration of
the biosphere long before we've managed to understand how it
works. It is calculated there are between 3 million and 10 million
species of plants and animals, of which only $1\frac{1}{2}$ million are
recorded, and very little is known even about these. The fact that
thousands of species will disappear by the turn of the century is
not just an academic irritation: our own survival depends on our

understanding of the intricate webs of life in which we're involved. Nor is it just a question of the 1,000 or so familiar and appealing mammals which are now threatened; our concern must be based on something more than the instinct to cuddle warm, furry creatures.

It's become apparent that there are two very different, but *not* mutually exclusive, schools of thought about this. On the one hand there are those who argue the 'enlightened self-interest' angle: that 'bio-impoverishment' is a threat to human welfare; that something like 40 per cent of modern medicines originated in the wild rather than the chemist's lab; that the shrinkage of gene pools is of particular concern to farmers and scientists at a time when only a very few seed varieties are being used, given that such genetically uniform crops are far more vulnerable to disease.

On the other hand, there are those who argue that this utilitarian approach is inadequate, and that non-economic, philosophical considerations should weigh as heavily as economic ones. Biologist David Ehrenfeld calls this the 'Noah Principle', claiming that 'long-standing existence in nature is deemed to carry with it an unimpeachable right to continued existence.'[3] That is obviously going a great deal too far for the authors of the *World Conservation Strategy* (WCS), who define conservation as 'management of the human use of the biosphere so that it may yield the greatest sustainable benefit to present generations whilst maintaining its potential to meet the need and aspirations of future generations'. You can't get much more utilitarian than that, and such a definition seems quite inadequate.

Are we to assume, for instance, that as soon as stocks of whales recover whaling will simply resume? One might have thought that in the preservation of the whale, ecological wisdom had won a signal triumph over narrow commercialism – if only just. And yet the existing controls imposed by the International Whaling Commission, including the prohibition of trade in most whale products, are still seen by some countries as purely temporary, designed to allow whale populations to recover. More fundamental ethical issues remain unresolved. Even though the economic benefits from whaling are slight, and substitutes can be found for all products, a complete ban on whaling for all time could still not be justified according to the WCS's definition of conservation.

Or are we just to make an exception of the whale? The UN Draft Plan on marine mammals posed the rhetorical question: 'What is the significance for management of the discovery of the whale song?' (that strange, haunting communication between these magnificent creatures). To an ecologist the answer is clear: where once we supposed that the saving of the whale was a symbolic turning point, the resumption of commercial whaling, however carefully 'managed', would symbolize the final and irretrievable abdication of our responsibility of stewardship on this planet. *That* is its significance.

The same dilemma is apparent in the controversy over vivisection. There are those who argue that the 83,000 animals put to such use every week in British laboratories are essential to 'progress' and human health. Ecologists argue that this utilitarian justification for cruelty to animals is not only ethically unacceptable, but utterly fallacious even in its own terms. The vast majority of experiments are quite unnecessary, often being needlessly duplicated on account of competition between different drug companies. Many new drugs are produced only to swell profits rather than to alleviate human suffering, and animal tests are by no means infallible indicators of the toxicity of certain products, as the example of Thalidomide demonstrated all too clearly. And what possible justification can there be for the use of animals in tests on cosmetics, or indeed in research into ailments that are uniquely human, such as smoking, alcoholism, stress and drug addiction? It is to us all part of an utterly irrational concept of 'progress' that animals should be blinded, poisoned, mutilated, electrocuted, irradiated and scalded, often without any anaesthetic, as an essential means of achieving it – all the more irrational in that there *are* alternatives that do not depend on the use of animals, but which government and industry between them refuse to develop. A purely utilitarian ethic is clearly quite inadequate in this and many other areas of modern life.

In the Name of Efficiency

Nowhere is this more clearly demonstrated than in contemporary farming practice. Farming is a biological activity, *not* an industry.

By turning agri*culture* into agri*business*, by operating according to an urban industrial model of economic efficiency, we have created some staggering problems for ourselves. And yet, to listen to the National Farmers' Union or to representatives of the farming 'lobby', ours is the most productive and efficient agricultural system in Europe. In one sense, it undoubtedly is: the area used for agriculture has been reduced by about 4 per cent since 1945; during the same time, the number of people working on the land has been drastically reduced, from 540,000 to 175,000 full-time workers, and yet the stock carried on it has increased by 60 per cent. The number of farms has been halved since 1945, and their average size has increased from about 80 to 150 acres. Output from less land and far fewer workers has therefore more than doubled. This increase in efficiency and productivity is usually referred to as the 'rationalization' of agriculture, and, so the argument goes, it is this that has allowed the consumer to benefit from cheaper food.

Those are the benefits, proclaimed the length and breadth of the country by most farmers and most politicians. The costs usually appear in the small print as the unavoidable consequences of achieving such 'progress'. We have already considered the extraordinary implications of a thriving agricultural system co-existing with the collapse of rural society; of greater import even than this is the damage done to the countryside through the adoption of capital-intensive, monocultural systems. Conservation is now seen as an unnecessary constraint on productivity, and many farmers either don't care or can't afford to care. They have resisted any attempt to correct the balance between the private right to profit and the public right to an unspoiled heritage. The countryside at large has *no* protection, for farmers are not subject to normal Town and Country Planning Acts, and are allowed to carry out developments without any planning permission.

In 1979, Marion Shoard's courageous book *Theft of the Countryside* first showed the full extent of the damage being done. This book had a dramatic impact on the whole countryside debate, not least because of her estimate that support for agriculture was costing the taxpayer and consumer around £5 billion a year. Even *The Times* was moved to declare that 'The system is insane.' In a more recent report, Friends of the Earth have

calculated that since 1949 Britain has lost 95 per cent of its herb
and wildflower meadows, 80 per cent of its chalk and limestone
grasslands, 50–60 per cent of its lowland heaths, 50 per cent of its
moorland grazing land, 30–50 per cent of its ancient woodland,
and 25 per cent of its hedgerows – 126,000 miles of them.[4]

The wholesale industrialization of agriculture has destroyed
many of our rarer flora and fauna. The Nature Conservancy
Council has described the losses as 'catastrophic'. Conservation
legislation has made almost no impact, and the 1981 Wildlife and
Countryside Act, which relies on voluntary restraint and cash
payments to persuade farmers not to develop on Sites of Special
Scientific Interest (SSSIs) or within our National Parks, has failed
totally to protect even those limited areas. SSSIs are now being
destroyed at a rate of about 15 per cent a year. It's not only the
well-known rarities that are now disappearing, but also many
species of plant and butterfly that were once considered common.
Charles Secrett, Friends of the Earth's wildlife campaigner, lists
the damage: four out of forty-one breeding species of dragonfly
extinct in recent years; all fifteen bat species now endangered;
nine out of twenty-five species of bumblebee at risk. Cowslips,
nightingales, otters and even primroses are increasingly rare
sights. It's a long and depressing list, headed by the natterjack
toad, which survives only because its only known site in the south
of England happens to be on land owned by the Ministry of
Defence – as Secrett says, 'under a security blanket that any of
Britain's 180 or so endangered species of animals, birds and plants
might well envy'.

A High Price for Cheap Food

One further aspect of agribusiness that affects millions of animals
is its dependence on factory farming. This too, we are told, has
become 'necessary' to provide the nation with cheap food, despite
the fact that the population has hardly risen at all since the 1950s
when the nation seemed to survive quite adequately without
factory-farming techniques. Our desire for cheap animal products
apparently justifies the extremely cruel conditions in which a
large proportion of chickens, pigs and calves are raised. As far as

the Ministry of Agriculture and Fisheries is concerned, high yields and high profits are infinitely more important than animal welfare. Time after time they assure us that there is no 'firm proof' that animals feel pain, and therefore no 'scientific basis' for discouraging factory farming.

It has apparently not yet crossed their narrow little minds that such practices are *not* particularly efficient if looked at from the point of view of the best possible use of protein. In terms of *energy* costs rather than *cash* costs (i.e. how much net energy we get out of the process for the energy we put in), the average Chinese peasant is far more efficient than the average British farmer. In a world of high resource costs and high unemployment the whole method by which we grow, process, package and distribute our food is desperately wasteful. Excessive packaging is responsible for generating hundreds of thousands of tons of waste. While oil was cheap, this could be accepted as more or less rational, although it was always energy-inefficient, but from now on, our dependence on capital-intensive, fossil-fuel farming means that the price of food must inevitably rise very sharply indeed.

As you can see, the costs are already mounting. But to present the full picture, one needs to draw up this particular balance sheet as one would the accounts of any other business. I am indebted to the ideas of Bob Waller for much of the foregoing analysis, and in particular to his excellent Green Alliance pamphlet *The Agricultural Balance Sheet*, which may be summarized as follows:

Costs
Decay of our rural communities
Elimination of the small farmer
Unemployment among farm
  workers
Destruction of the countryside
Extensive loss of flora and fauna
Huge import bills for fertilizers
  and feedstuffs
Dependence on cheap oil
Dependence on pesticides and
  other chemicals
Unnecessary cruelty of factory

Benefits
Higher productivity
Greater efficiency
Cheaper food in the
  short term

farming
Decline of husbandry and farming
    skills
Misuse of the soil through mono
    cultural farming
Wasteful surpluses and food
    mountains
Higher processing, packaging and
    transport costs
Misuse of subsidies and tax
    concessions
Concentration of power and
    wealth in fewer hands
Dearer food in the long term
Poorer-quality food and lower
    standards of nutrition

Realistic cost/benefit analysis as regards modern farming makes a complete nonsense of words like 'productive', and 'efficient'. Farmers are themselves as much the victims of this approach as its perpetrators; even when they know that what they are doing contradicts their own good sense and natural wisdom, economic 'necessity' compels them to deny such instincts. Some are more reluctant to comply than others, but we are rapidly approaching the time when we shall no longer be able to trust farmers to act as stewards of the land. As Bob Waller says, 'In the present phase of history, the technological imperative has triumphed over the biological needs of the land and animals, and the psychological and spiritual needs of man.' The resulting confusion is all part and parcel of a much broader spiritual malaise: 'Farming is a biological activity, not a mechanistic one; land, plants and animals are living organisms, not machines.'[5]

The World as Machine

To understand this confusion (and understand it we must, for otherwise there can be no redress against the power of industrialism), it is necessary to go back about three hundred years, when a complete shift occurred in our relationship with the rest of creation. In 1686, Isaac Newton published his *Mathematical Principles of Natural Philosophy*, in which he described the world as a kind of machine, with him working away as a mechanic to discover just how the machine operated. That's what is meant by a 'mechanistic' view of life, and it is one shared by many both before and after Isaac Newton.

The development of this philosophy gradually drew people away from contact with the natural world. Nature came to be seen as a 'commodity' to be bargained for, a 'utility' to be exploited. People began to talk in terms of confrontation, with 'mankind opposed to nature'. Man had been set 'in dominion' over all other creatures, and part of our relationship with God was to exercise this dominion on His behalf. Over the next couple of centuries we went a step further and decided that it would be a good idea to dispense with God altogether, substituting in the place of religion a purely mechanistic interpretation of the origins and workings of the planet. Since the time of the Industrial Revolution, the dominant philosophy throughout Europe has been that of 'scientific materialism': all explanations of life are reduced to the material: what cannot be scientifically proved cannot exist; what cannot be measured cannot matter.

This impersonal, rationalizing philosophy, with its emphasis on purely economic and scientific values and progress, has underpinned the development of both capitalism *and* communism. On both sides of this ideological divide, people have been motivated by resentment of religion in the belief that anything metaphysical is likely to lead people astray by distracting them from the physical realities of life. Marx, the creator of 'scientific socialism', ensured that right from the start the collectivist forces of socialism were allied to the ruthlessly sceptical approach of scientific materialism. All religious and mystical experiences were from then on adjudged to be counter-revolutionary.

As Roszak points out, there is an irony in both the collectivist and individualist traditions that defeats their highest purpose. In reducing all explanations of life to the material, they have attempted to wipe out the supernatural, and, in so doing, to wipe out an important dimension of human experience. I don't claim to be any great expert in working out what makes people happy, but it strikes me that a life without some kind of supernatural or spiritual or mystical dimension is unlikely to be a full life. Our culture has completely lost that sense of balance, and our spiritual dimension is often contemptuously dismissed as 'purely irrational'. But the industrial ethic is no more likely to succeed in suppressing the spiritual dimension than the Middle Ages were in suppressing rational and scientific enlightenment. In reaction to a narrow, oppressive form of medieval Christianity, people at that time turned away from contemplation of the inner life to celebration of the outer life. It seems reasonable to suppose that a similar reaction may be about to set in against the equally narrow and oppressive materialism with which we are now burdened.

Mindless Materialism

There is no doubt that the attempt to do away with metaphysics was for the most part carried through by people with the best and purest motives; but in their belief that the better side of human nature would flourish in a secular world as it had never flourished in a spiritual world, they have been utterly confounded. Of their inspired humanitarianism we have managed to make an uncaring, life-destroying metaphysics all of its own.

For ecologists, struggling to make sense of a world that seems intent on a particularly sordid kind of self-destruction, there are several consequences we have to deal with. The first is the unconsciously held assumption that instead of humankind being *a part of* nature, we are somehow set *apart from* it. People have suppressed that vital link between themselves and the Earth – so much so that when you talk of 'harmony between ourselves and all living creatures', they reckon you've probably spent too much time talking to the trees! Apart from the occasional spasm of wonder inspired by superb natural history programmes on television, the natural world remains something alien, something to be mechanically exploited by science and technology.

Secondly, as I have tried to demonstrate throughout this part of the book, ecologists are up against the mind-boggling irrationality of contemporary reason. 'Rationalization', 'maximizing efficiency', 'economic necessity': these have been the passwords of

our industrial culture, and much good they have done us. By
repressing feelings, sentiment, intuition, we have ended up with
an incomplete, degraded and horribly aggressive image of human
nature; by denying the creativity and 'transcendent aspirations' of
people, we have become dependent upon a sick caricature of
reason and science. 'Long before the barricading of the streets
begin, the spirit of man starts barricading itself, and then proceeds
to throw stones through the windows of the mind.'[6]

And, last, we are up against materialism, the deeply ingrained
belief that the pursuit of material wealth is an end in itself, and
that there can be no salvation or meaning to life except through
material fulfilment. It has been one of society's most strongly held
dogmas that all we need to do is increase our net wealth,
distribute it more justly, and a prosperous and contented society
will be the automatic consequence. As René Dubos points out in
The God Within, the only successful myth that the West has
created on its own behalf is that of Faust, who signed away his
soul to the devil in return for wealth and power here on Earth. It
just so happens in Goethe's play that Faust is saved by the power
of love and is received into Heaven; in terms of our particular deal
with the devil, who can say what the source of redemption will
be? Certainly not our politicians or economists, as they dance ever
more frenetically to the tune of mass consumption.

Our whole philosophy is both spiritually and ecologically
unsustainable. 'We have worshipped the god Economos; and it
has given us material plenty. But it is unable to give us the
meaning of life.'[7] These words of Henryk Skolimowski's bring us
back full circle to the thoughts of Fritz Schumacher with which
this chapter opened. He believed that ours is primarily a spiritual
crisis, a form of sickness for which we continue to take the wrong
medicine, and criticized those who hold 'a view of the world as a
wasteland, in which there is no meaning or purpose, in which
man's consciousness is an unfortunate cosmic accident, in which
anguish and despair are the only final realities'.[8] Ecologists must
endeavour to break this cycle of cynicism and spiritual despair, for
within such a framework it remains impossible to establish any
other world view than that of industrialism.

It's no easy thing for a politician to stand out against the tide of
materialism, for as a politician it is natural to link one's interests

and ambitions with the apparent self-interest of one's voters. The consequence is chronic 'Jonesism', as Hazel Henderson dubs this constant need to keep up with the Joneses. Psychologists tell us that the very process of acquisition and consumption of material goods is for many a validation of their own significance, even of their very existence. If that is the case, we are in real trouble. In a world of growing scarcity and resource constraints, conspicuous consumption is hardly the most convenient or intelligent way for people to establish their identity. The single-minded pursuit of wealth, spurred on by the insidious appeals of mass advertising, recognizes no inherent limitations. There is no point at which people seem to agree that enough is enough. Yet the environment that has to cater for this insatiable drive is strictly limited. So we must return to the balance sheet, and a final assessment of the benefits we get and the costs we pay as a result of our way of life.

Fudging the Figures

It must be clear by now that ecologists reject the very premises on which any purely economic cost/benefit analysis is based. Many things simply can't be quantified. By forcing non-economic values into the framework of a specifically economic analysis, everything is reduced to the lowest common denominator. The priceless cannot be given a price; the immeasurable cannot be measured. To do so can only lead to deception, for if everything can be reduced to this sort of narrow economic analysis, then we have accepted that money values are the final arbiters in all debates.

Issues of vital importance are cunningly 'depoliticized' by being presented in this way, and genuine political debate becomes impossible if all opposition is dismissed as 'irrational' simply because *its* rationality cannot and will not be reduced to so deceptive an analysis. What is at stake is not just the economics of pollution or nuclear power or modern farming, but the under-lying values and conflicting views about our moral and social order. Moreover, strict cost/benefit analysis is extremely ineffi-cient when it comes to dealing with what are called 'opportunity costs'. Every usage of time or resources requires the sacrifice of the *alternative* uses to which these could have been put. When you sit slumped in front of the TV on a Saturday night, your opportunity costs include everything else that you could, and possibly should, be doing. When this Government decrees that *90 per cent* of all money spent on research into and development of new energy sources should be spent on nuclear power, the opportunity costs include what *might* result if we spent an equivalent sum on renewable sources. With each and every one of the problems we've considered in this section, there comes a point when any conventional cost/benefit analysis simply breaks down.

The irony for ecologists is that although we reject the very process, we reckon the argument can usually be won even in conventionally narrow, economic terms. Consider the example of high military spending: the benefits of such spending are presumed to be increased security and 'considerable economic advantages'; on the other side of the balance sheet we must consider both resource costs and opportunity costs. A considerable proportion of the resources extracted from the Earth are used to make weapons: the military use of oil amounts to more than 5 per cent of the global total; more aluminium, copper, nickel and platinum are used for military purposes than are used for *all* purposes in Africa, Asia and Latin America combined. We are literally consuming the planet in order to make it safe. Opportunity costs include all civilian goods and services foregone as a result of military spending, including the colossal waste of skilled scientists and the misuse of other human talents. Add to this the case already presented, that far from being a source of economic benefits, high military spending carries high economic penalities, and one can appreciate that even in conventional terms, this Government's policy is grotesquely irrational.

At this point ecologists might well throw up their hands in despair, because throughout this whole laborious process we've missed the real point – namely, that it can no longer be proved that increases in military spending necessarily increase the security of one's country, and it may indeed be demonstrated that they undermine it. This is a question of values and not of economic costs. One can fudge the issue by saying that the economic 'cost' of maintaining a state of perpetual war readiness is in itself so serious a factor as to render war far more likely, over and above the stockpiling of weapons that goes on in the process. But at a deeper level, what the greens are saying is that we are already *at war* because of the way in which we choose to define individual and national security. We are at war with each other, since 'success' in today's materialistic world is possible only at someone else's expense; we are war with other nations because our narrow, chauvinistic concept of national sovereignty defies any possibility of building a new internationalism; and we are at war with the planet, since today's affluence is achieved only at the expense of our natural wealth. Just try costing out that little lot and you will see how ridiculous a process it becomes.

Earthrights

My contention is that we are paying too high a price for our Faustian bargain. We have indeed seen extraordinary progress since the time of the Industrial Revolution, but not without incurring very considerable ecological debts which we are only now having to pay off. In the process we have lost our sense of balance, and seem incapable of distinguishing between genuine reason and facile rationalization. When our vision shrinks, our reason shrinks with it. In essence, the green critique of industrialism argues that we have lost our way by disregarding the planet and by disregarding the spirit. We have forgotten our dependence on the biosphere, and we have suppressed the gentler, deeper side of human nature. We have lost touch with the Earth and with our real selves.

As a result, we have ended up in a right old mess. Worse still, we are taking decisions now which will guarantee even more of a mess for future generations. The waste generated by our nuclear power programme has become a symbol of the extent to which we are prepared to commit future generations to share in our follies. There is no 'solution' to the problem of nuclear waste, nor can there ever be except to stop producing it in the first place. All materials eventually decay; a nuclear waste container will not endure even as long as the Pyramids, though its radioactive contents will remain a threat for far longer than that. It's bad enough right now for the inhabitants of Billingham or Elstow or any other spot earmarked as a site for nuclear waste disposal, but spare a thought for our descendants who can hardly be reckoned accountable for our decisions, though they will certainly be paying the price for them.

If we are to leave the planet as we would like to have found it, and thereby to liberate the future from the problems and so-called 'solutions' of the present, we must first recognize that ecology and good economic management are now one and the same thing. Every threat to the Earth is a threat to ourselves; every wound inflicted on the Earth is our own wound. Wealth and welfare simply cannot exist in a world that is ecologically unhealthy, and the production of wealth can no longer be separated from the conservation of the source of that wealth. When I first read Theodore Roszak's *Person/Planet* it was as if the whole thing was suddenly illuminated for me. The links between people and the planet, and the understanding that real self-interest has to be articulated in terms of those links, had previously remained obstinately obscure. Roszak's purpose was

> to suggest that the environmental anguish of the Earth has entered our lives as a radical transformation of human identity. The needs of the planet and the needs of the person have become one, and together they have begun to act upon the central institutions of our society with a force that is profoundly subversive, but which carries within it it promise of cultural renewal.'[9]

Part of such a process will be to develop a new sort of radical politics that opens itself to the spiritual dimension of life, and a new approach to 'accounting' which acknowledges both economic *and* non-economic values, liberating the creative potential of individuals and whole societies, showing equal reverence for one's own life, the life of others and the Earth itself. Such a balance sheet will not be easy to draw up. But that, quite simply, is what all those involved in green politics are endeavouring to do. 'The rights of the person are the rights of the planet.'[10] It is to those rights that we must now turn.

Part Three

Turning to Green

9
Ecologics

Enlightened Self-Interest

The fact that people's rights are being denied is in itself a serious enough problem. The fact that the majority of people are not aware of this makes it even more serious. And the fact that there are so few, apart from the greens, who are prepared either to inform people of the denial of their rights, or to help them to fight for those rights, turns a problem of indifference into a crisis of inaction. It is ironic that it should be the ecologists, whose politics have hitherto been kept at a somewhat elevated level, who are now going to have to demonstrate that even at the level of the lowest common denominator (namely, straight self-interest) people are still getting a pretty bad deal.

There was a time when ecology was dismissed as a middle-class fad, particularly by those whose vision had been shrunk by years of left-wing pseudo-radicalism. It was treated as a kind of 'supplementary benefit', to be attended to when all the *real* problems had been solved, and in the meantime it did nobody any harm that the slightly eccentric middle classes should exercise their social conscience on the issue of the day. How grotesque a misapprehension that was – and how seriously it has set back the possibilities of developing a genuinely radical opposition to today's industrialism.

It is really not saying very much to claim that the majority of greens are well-educated and tend towards certain middle-class occupations. So are and do the majority of industrialists. And yet I would claim that the divide between greens and industrialists, with their opposing world views, is of far greater significance than the notional class distinctions by which most politicians are still obsessed. Yet one thing *is* clear: even if we continue to think in terms of working class and middle class, it is not the latter that has most to worry about in terms of the current crisis. It is the middle classes that have the flexibility to weather traumatic shifts in social and economic patterns; by and large, they are not the ones to suffer most from mindless jobs, dangerous working conditions, a filthy, polluted environment, shattered communities, the exploitation of mass culture, the inhumanity of bureaucrats and the mendacity of politicians. Given such circumstances, one must of course acknowledge that the post-industrial revolution is likely to be pioneered by middle-class people. The reasons are simple: such people not only have more chance of working out where their own *genuine* self-interest lies, but they also have the flexibility and security to act upon such insights.

It seems unlikely that most people will be able to discern their own genuine self-interest until our society begins to shrug off the curse of individualism. There may well have been a time, at the start of the Industrial Revolution, when Adam Smith's assertion that the sum of individual decisions in pursuit of self-interest added up to a pretty fair approximation of public welfare, with the 'invisible hand' of the market ensuring that individualism and the general interest of society were one and the same thing. But in today's crowded, interdependent world, these same individualistic tendencies are beginning to destroy our general interest and thereby harm us all. They create needless conflict; they undermine our biological support systems; they exacerbate loneliness and alienation.

By emphasizing the least attractive qualities of human nature – our competitiveness, our greed and our selfishness – we have disregarded the fact that *rights* must always be balanced by *responsibilities*. We have therefore ended up in the peculiar position where we say to ourselves that we can 'afford' every conceivable modern convenience from videos to fast food, and that we can 'afford' billion-pound cosmetics and 'personal hygiene' industries, but that we can't afford nurses, teachers or social workers. The invisible hand of the market, working exclusively through individual preferences, cannot deal with these public-sector needs, let alone the need to protect our common wealth and common resources. Garrett Hardin's much quoted *The Tragedy of the Commons* remains the simplest explanation of this fundamental truth. In pre-industrial Britain, farmers grazed their sheep and cattle on the commons, and everyone benefited from such a sharing. But imagine if one or two farmers suddenly realized that they could maximize their own individual interests by grazing more animals than their neighbours. To begin with things would be fine, but suppose others noticed what was going on and realized they had better do likewise? In no time at all the commons would be destroyed through overgrazing. And then who benefits?

Individualism alone cannot provide people with a yardstick for their own genuine self-interest. For what may appear to be in their interests turns out to be something that destroys those interests. We have already seen how false accounting procedures have begun to threaten the maintenance of our fragile biosphere. And yet isn't it blindingly obvious that in caring for the planet, we are actually caring for ourselves, protecting our own self-interest? We've already seen that distorted notions of work and community are undermining the fabric of our society through wholesale alienation. And yet isn't it obvious that in caring for others, for our communities, for the work we do, we are in fact caring for ourselves?

I do not believe that the majority of people will change until they believe it is in their own interests to do so. That may be a somewhat cynical position to adopt, but minority politics plays havoc with ideals. A reinterpretation of enlightened self-interest is therefore the key to any radical transformation. And that is why we must argue our corner from within the heartland of conventional politics, for it is the politicians of today who make it so difficult for people to see where their real interest lies. Some through ignorance, some through weakness, and some through deliberate dishonesty, they all play their part in promoting the life-denying illusions of a fading industrial dream – a dream that is now incapable of meeting the needs of more than a tiny minority.

The Breakdown of Dissent

We have already considered the dangers of technological determinism. They are nothing compared with the blight of political determinism. For a brief while Mrs Thatcher was nicknamed 'Tina', on account of her manic reiteration of the catchphrase, 'There is no alternative.' The opposition parties naturally got a bit irritated by this, as they all believe that they *do* have an alternative. Yet an analysis of industrialism reveals that economic necessity is the motor that drives *all* political machines; the 'imperative' of maximizing production and consumption has in fact throttled the alternatives. They have not yet learned that there is only one imperative, and that is the ecological imperative of learning to live in harmony with the planet. All the rest are just rationalizations of politicians' power games.

The tragedy is that almost all the voices of so-called 'dissent' have gradually been sucked into this nexus of non-opposition. Academics, the media, even the established Church, they all bend the knee at the right place and the right time. The left are no better than the right, and the centre is the worst of the lot. But it is perhaps the trade unions that epitomize this breakdown of dissent. In many ways they have become the rather undignified mirror image of the world of industrial capitalism. Stephen Cotgrove's survey indicated that there is great concern among the unions about many environmental and social issues, such as better working conditions, tighter controls on pollution, better procedures for participation, a more personal and humane society. But such concerns do not, it appears, diminish their continuing, rock-solid support for today's fundamental economic goals. As Cotgrove says, 'They are remarkable for their strong support for both material and post-material goals.'[1] In common parlance that's known as having your cake and eating it, and it can't be done. As representatives of 'working-class interests', they remain bogged down in the arguments, hopes and fears of industrialism;

they do not see the new powerlessness that grips society, of which they themselves are a tragic part.

There's no such equivocation on the part of the media. They are the transmitters and reinforcers of the values of industrialism, and television is the worst offender of the lot. As a teacher I knew only too well how much more powerful television is than schools, churches or even families in terms of creating a set of values or 'consensus' opinions. Day after day one is exposed to a totally distorted portrayal of people, concentrating almost exclusively on the greedy, the violent, the superficial, the selfish and the apathetic. At a time when we require the very best, we are fed a diet of the very worst. Except for the occasional daring documentary, current affairs programmes rarely do more than scratch the surface of today's problems. Worst of all, television has become a tool of our mass-consumption society, reinforcing materialistic attitudes and wasteful habits. The commercial TV companies make their profits by delivering the largest possible audience of potential consumers into the avaricious grasp of the advertisers. This audience is bombarded with pleas and exhortations to consume more, and though the BBC is not obliged to prostitute itself in this way, it is obliged to compete for that same mass audience by propagating the same mass values. Duane Elgin, the American author, puts it like this: 'We have inadvertently handed over a substantial portion of our cultural consciousness to an electronic dictatorship that promotes extravagant consumption, social passivity, and personal impotence.'[2]

We face an appallingly difficult period of transition as we move towards a more sustainable society. It will require the most massive adult education programme ever imagined, in which television, by far the most powerful communications medium, will have to play the major role. A democracy cannot function properly without an informed and participating citizenry. Whether television merely reflects the prevailing social ethos, as it claims, or whether it is instrumental in shaping it, it is quite incapable at the moment of carrying out any such role. To remedy that sad state of affairs is partly the responsibility of those who work in television; but primarily it is ours, for it is we who must redefine the 'public interest' that television is presumed to serve.

Growing Pains

When it comes to definitions, politics certainly plays havoc with the English language, misusing and distorting certain words so that they end up meaning the very opposite of what they should mean. So 'rationality' has come to represent a measure of the absurd, 'productivity' a measure of destructiveness, and 'progress' a measure of just how quickly we can march backwards into the future. Consider for a moment the fuss and nonsense about 'growth'. The most convenient method of comparing the economic performance of different countries is to consider their rates of economic growth; when a country moves up the international league table, chests swell, but when it moves down, there's the most ghastly gnashing of teeth. But how is it that people expect rates of growth to go on growing? Every annual increase is related to the existing base of the economy, and since that's increasing every year, to look for increased rates of growth from an ever-increasing economic base is plain silly.

It is surely self-evident that we cannot continue expanding at past rates of growth, and yet since the war we've made this one measurement the ultimate arbiter of social progress. Because of our opposition to this manifest absurdity, ecologists are always seen as the 'no-growth party', the 'zero-growthers', though such a position is obviously just as absurd as that adopted by the 'infinite-growthers'. It is our contention that there will always continue to be *some* economic growth: in the developed world, there will be limited growth in certain sectors of the economy, even though the overall base will no longer be expanding; in the Third World, there will have to be substantial economic growth for some time, though with much greater discrimination as regards the nature and quality of that growth. All economic growth in the future must be sustainable: that is to say, it must operate within and not beyond the finite limits of the planet. On top of that, the emphasis will not fall exclusively on growth in tangible, quantitative possessions, but will deal equally with growth in personal and human resources. Increasing the skills and knowledge of people is one form of growth that is not constrained by the finite physical limits of our planet.

Such changes will mean doing away entirely with the nonsense of *Gross* National Product – for once a word is being used correctly! Even conventional economists admit that the hey-day of GNP is over, for the simple reason that as a measure of progress, it's more or less useless. GNP measures the lot, *all* the goods and services produced in the money economy. Many of those goods and services are not beneficial to people, but rather a measure of just how much is going wrong: increased spending on crime, on pollution, on the many human casualties of our society; increased spending because of waste or planned obsolescence; increased spending because of growing bureaucracies. It's all counted in, as if these were positive contributions to wealth. It's hardly surprising, therefore, that as GNP rises, it doesn't necessarily mean that either wealth or welfare is increasing proportionately. Indeed, there are some economists who contend that the only aspects of GNP that are consistently rising are the very social and environmental *costs* to which I've referred.

There have recently been one or two attempts to find alternatives which *deduct* the social and environmental costs rather than adding them in. The Japanese system of 'Net National Welfare' does this. The US Overseas Development Council has for a long time been using the Physical Quality of Life Indicator (PQLI), based on the factors of life expectancy, infant mortality and literacy. Other alternatives have incorporated the value of the household or domestic economy, or even the percentage of national energy use that is derived from renewable sources; one report in Canada, on the subject of a proposed Quality of Life Indicator, suggested that thirty-seven factors should be taken into account, including physical and psychological security, personal dignity and 'self-actualization'! We're back full circle to the fallacy of trying to measure the immeasurable.

Paradoxically, progress in the future may consist in finding ways of *reducing* GNP. Today's planned obsolescence (whereby things are built specifically *not* to last so that new ones have to be purchased to replace them) raises GNP; production for durability (whereby things are made specifically so that they *do* last) lowers it. The wasteful consumption of non-renewable resources, particularly oil and gas, raises GNP; the development of renewable energy sources will drastically lower it, since fewer people will have so much to pay. Our understanding of 'efficiency' must

necessarily change in an age where we have a surplus of human
beings and a shortage of non-renewable resources, for only in a
labour-intensive society will it be possible both to conserve the
Earth's resources and ensure that the wealth of human resources
are expended for the benefit of fellow humans. I can remember
laughing out loud when I first read Ivan Illich's masterly account
of relative transport efficiency in the USA and so-called primitive
countries:

> The typical American male devotes more than 1,600 hours a year
> to his car. He sits in it while it goes and while it stands idling. He
> parks it and searches for it. He earns the money to put down on it
> and to meet the monthly instalments. He works to pay for petrol,
> tolls, insurance, taxes and tickets. He spends four of his sixteen
> waking hours on the road or gathering his resources for it. The
> model American puts in 1,600 hours to get 7,500 miles: less than 5
> miles per hour. In countries deprived of a transportation industry,
> people manage to do the same, walking wherever they want to
> go, and they allocate only 3 to 8 per cent of their society's time
> budget to traffic instead of 28 per cent.[3]

Managing our Wealth

In so different a world, the very notion of 'economics' will itself
be up for grabs. If we're being strictly accurate, this word too is
sorely misused: it has the same Greek root as 'ecology' (namely,
oikos, a house) and the word *nomos* means to manage. Theologians
used to use the word *oeconomia* to explain God's dispensations here
on Earth, but since the Middle Ages the standard of management
has declined drastically. Mrs Thatcher's very keen on presenting
an image as the 'housewife Prime Minister', but what housewife
or househusband would so 'manage' affairs that they not only
squandered their little nest-egg, but irretrievably soiled their little
nest in the process? We're about as good at managing the planet as
Billy Bunter was at managing his tuck-box.

In daily usage, the word 'economic' is like a magic password.
As long as something's 'economic', it must be all right; and if you
want to put down someone's good idea, just shake your head at
them and tell them it's 'uneconomic'. If something's ugly,

wasteful or unpleasant, it doesn't matter as long as it's 'economic'; and anything that doesn't yield an instant financial return must be 'uneconomic'. So it's uneconomic to think about the future, but economic to tear down all the forests. It's uneconomic to recycle things, yet economic to exploit the soil till it's barren. Modern societies obviously need some measure of profitability, but let us not confuse this with real *wealth*, and let us not adopt it as the ultimate arbiter of all we do and aspire to do. Oscar Wilde described a cynic as someone who knows the price of everything and the value of nothing. Ours is a cynical world, with a highly cynical interpretation of wealth.

In *The Sane Alternative*, James Robertson points out that the 'wealth of the country' usually refers to the activities of industry and commerce; it's something which they produce so that it can be spent on something entirely different, namely social welfare. And it doesn't much matter what they produce: laser death rays (real or toy), toxic pesticides, electric boiled-egg openers, cigarettes, cosmetics for cats – why discriminate when it's all 'creating wealth'? On the other hand, ordinary people doing ordinary things are using up that wealth, and the contribution which millions of people make to the common wealth of this country as teachers, doctors, nurses, housewives, social workers, prison visitors, charity volunteers is somehow not held to be *real* wealth at all.

At the individual level, 'wealth' means the visible symbols of affluence. It means consumer durables and credit cards and being rich enough to have a huge overdraft. How, oh how this is going to change! In a sustainable, ecological future, the wealthy will be those who have the independence and the education to enhance the real quality of their lives; the poor will be those who look back to an age where money might, but never quite did, buy happiness. The wealthy will be those who live and work in a friendly, mutually supportive community; the poor will still be trucking off to the cities in overcrowded commuter trains to do jobs they can't stand anyway. The wealthy will be those who make more of their own entertainment in a more convivial society; the poor will be twiddling the buttons on their cable TV videos trying to find the right brand of oblivion. The wealthy will be growing as much of their own food as they can, and growing it organically; the poor will be paying through the nose for an adulterated mess of potage. The wealthy will be re-using and recycling and taking pride in how long things last and how easy they are to repair; the poor will be wondering when the novelty went out of novelty. The wealthy will be fully involved in their parish or neighbourhood council, getting things done for themselves and their community; the poor will still be blaming the Government. Wealth, in both its physical and its spiritual dimension, will have regained its meaning. R. H. Tawney's words are as true today as when he wrote them:

> The most obvious facts are most easily forgotten. Both the existing economic order and too many of the projects advanced for reconstructing it break down through their neglect of the truism that, since even quite common men have souls, no increase in material wealth will compensate them for arrangements which insult their self-respect and impair their freedom.[4]

The Economics of Enough

The next chapter deals with green economics: economics as if people *and* planet really mattered. But I should warn the reader at this stage that you will find in it no reassuring panaceas, and none of the illusory promises with which politicians normally dress up their economics. The plain fact is that in terms of crude material wealth, we're not likely to get any wealthier. But the crude, quantitative, undiscriminating measurement of wealth is something of the past; what matters now is the quality of wealth.

The key distinction here is to be found in the difference between needs and wants. As Gandhi said: 'The Earth provides enough for everyone's needs, but not for everyone's greed.' The dividing line between wants and needs is a hard one to draw, but not drawing it will make things even harder. And yet the notion of human wants as essentially unlimited, given the right kind of technological breakthroughs, is the starting point for many an economic analysis. We cannot really envisage a new age unless we realize that some of our wants and so-called 'needs' are quite spurious, in as much as the satisfaction of them does not add much to individual or social welfare. Moreover, it may well be the case that the satisfying of such needs in one country means depriving people of the chance to meet their genuine needs in another country. 'What were luxuries to our ancestors have become necessities to us' is a poor rule of thumb for the modern age; for, as we have seen, in a consumer-oriented society no judgement has to be made of the nature or quality of production. The fact that many of the needs created or stimulated through the general pressure of materialism and the specific thrusts of mass advertising subsequently turn out to be artificial is of no concern to either politicians or economists. 'You are what you consume' is the advertiser's fiction, and it's one that all society endorses in the name of economic success.

In 'Peace and Permanance', the essay of Schumacher's I referred to at the start of the last chapter, he writes: 'The cultivation and expansion of needs is the antithesis of wisdom. It is also the antithesis of freedom and peace. Every increase of needs tends to increase one's dependence on outside forces over which one cannot have control.'[5] We can and must change some of our habits. As far as the greens are concerned, economics simply means managing our affairs in such a way as to meet our genuine needs. That means doing right by ourselves, and doing right by the planet. The economics of more and more, on which industrialism once thrived, and through which many once bene-fited, is no longer able to meet our genuine needs while at the same time protecting the rights of the planet and the rights of future generations. It's time for the economics of enough.

IO
Green Economics

Sustainability

Green economics is all about *sustainability* and *social justice*: finding
and sustaining such means of creating wealth as will allow us to
meet the genuine needs of all people without damaging our
fragile biosphere. It implies a straight choice between what we
have now (a consumer economy) and what we will need in the
near future (a conserver economy). It is no longer possible to
manufacture abundance through making unsustainable demands
on the world's resources and environment; we must therefore
substitute more appropriate patterns of consumption that will
make for wiser use of both the world's resources and the human
resources at our disposal. The Green Party's booklet on
employment, *Working for a Future*, looks forward to such an
economy in these terms:

> In the long run, all nations will *have* to learn how to manage the
> demands of their people in a stable-state economy. The character-
> istics of such an economy are clear: reduced industrial throughput,
> greater self-reliance and sustainability through largely
> decentralized economic activity, maximized use of renewable
> resources and conservation of non-renewable resources, a far-
> reaching redistribution of wealth, land and the means of produc-
> tion, with the possibility of more fulfilling, personally satisfying
> work, all set within a more co-operatively based framework, and
> enhanced by the use of new technologies where they complement
> the above features.'[1]

As I've said, the transition is bound to be difficult. Attitudes, values and lifestyles will all have to adapt as best they can as we reach out into new territory. Such a process amounts to a revolution in all but name, and as such will touch people's lives in a most personal and direct way. Economics is not primarily about rates of growth, or tinkering with the money supply, or any other vague macroeconomic abstractions: it's about the work we do, and the rewards we get from that work. And that's where the revolution starts.

Greening the Work Ethic

I must confess to being revolutionary in a very old-fashioned way when it comes to work. The statement of Thomas Aquinas, 'There can be no joy of life without the joy of work', just about sums it up for me. I'm one of those who consider work to be a necessity of the human condition, a defining characteristic of the sort of people we are. I'm not at all surprised, in a world of mindless industrialism, that work is little more than 'sorrowful drudgery' to so many people, but to my mind that does not suggest we should try to eliminate work altogether, but rather that we should *liberate* it. Far from universal automation 'solving our economic problems', I believe it would so impair our humanity as to make life utterly meaningless. The active participation of people in the work of their society, rather than their displacement from it, strikes me as a precondition for the development of any sane, sustainable society.

As we saw in chapter 6, the notion of 'full employment' in the conventional sense no longer has any relevance; indeed, it now serves as a useful hallmark of the irrelevance and even downright dishonesty of those benighted politicians who still promise it. Even were it possible, it still fails at the first hurdle: it does not distinguish between employment as a nine-to-five drudgery job, a form of 'wage labour' which has to be suffered as the only means of maintaining one's standard of living, and employment as the effective utilization of a person's skills and time, providing sufficient material reward to allow him or her to live in reasonable comfort. *Good work*, which creates sustainable wealth and which redistributes it in comparatively direct ways, is the cornerstone of a healthy economy. Ecologists' ideal definition of work is therefore that it should:

> provide a reasonable standard of living;
> be personally satisfying;
> be environmentally sound;
> be socially useful;
> be available for everyone who wants it.

In a society that operated according to such an ethic the distinctions between work and leisure would obviously be reduced, for the two would become complementary aspects of the same process. Through *both* people should be enabled to recognize the personal value of the way they spend their waking hours, and to assert their freedom of choice in developing their own potential. Work, whether paid or unpaid, should offer everyone the opportunity of fulfilment in confirming his or her own worth and relationships with others. People who experience such fulfilment today are said to have a 'vocation' or calling. How strange it is that this should have become so specialized, almost elitist, a word, as if that calling to perform *responsible* work were only there in a tiny number of people. Roszak speculates on the power that would be released into the world if everyone had a chance to involve his or her whole personality in work, regardless of its profitability or productivity or cost/benefit credentials.

'Unrealistic', do I hear you say, or 'uneconomic' even? Not so! Ever since the start of the Industrial Revolution, politicians and economists have worried themselves silly about supply and demand, productivity levels, the balance of payments, government spending or the international debt. The emphasis has always been on the *product* and the money involved; they have scorned the *process* and the people who make it possible. The consequences of this are only now apparent, for it is no longer realistic to go on as we do now, ignoring basic human needs and deliberately destroying the planet in the process.

So Who's a Luddite?

Our whole attitude towards technology must change accordingly. With more people and fewer resources, the capital/labour ratio must start shifting back towards labour-intensive production; as the price of ever-scarcer and more expensive inputs increases, so it must eventually become 'economic' to end the process of labour displacement in many areas of the economy – thus re-establishing the proper balance between capital and human inputs. It is, as so often, the balance that is lacking, and that, I suspect, is why many greens feel a certain sympathy for old Ned Ludd, despite the fact they don't much relish being called 'neo-Luddites'. The Luddites, who became prominent in the early nineteenth century, are much maligned, for their prime concern was to ensure the sensitive adaptation of industrial technology to the patterns of their craft. They only started smashing up the new looms when they realized that these were being used as a weapon to destroy their rights as skilled workers, and that they were being excluded from having any say in just how this technology should be introduced. Lewis Mumford took up their case from the other side:

> What shall we say of the *counter-Luddites*, the systematic craft-wreckers, the ruthless enterprisers who, during the last two centuries, have in effect confiscated the tools, destroyed the independent work-shops, wiped out the independent tradition of handicraft culture... and sacrificed human autonomy and variety to a system of centralized control that becomes increasingly automatic and compulsive?[2]

Ecologists are *not* hostile to technology *per se*, and the use of advanced technologies of many kinds is essential to the development of an ecological society. One undisputed advantage of microprocessors is that they will enable machines to do jobs that are boring, unhealthy, unpleasant and dangerous; there should therefore be a considerable improvement in the working environment as a result. Moreover, much of the new technology could facilitate production of a less energy- and resource-wasteful nature. Benign technologies that serve human needs and remain firmly under people's control, and the sort of advanced, inexpensive equipment that is available for small-scale work in many fields, will do much to ease the transition.

It is a matter of choice whether technology works to the benefit of people or perpetuates certain problems, whether it provides greater equity and freedom of choice or merely intensifies the worst aspects of our industrial society. It is quite clear that in today's economy the introduction of labour-saving machinery strengthens the hand of a very small group of managers and technicians representing the interests of big business; without some acceptance of the need to control the development of technology, one is merely confirming the maintenance of the status quo, and thereby endorsing the inhumanity, the inequality and the fundamentally exploitative politics on which the status quo rests.

There must therefore be some control if the indiscriminate application of new technology at the whim of so-called 'market forces' is not to cause terrible damage. It will first be necessary to work out some guidelines (e.g. who benefits? Is it in the interests of the workers and community as a whole? Does it harm the environment? Is it a sensible use of scarce resources? Does it improve the overall 'quality of life'?), while acknowledging here and now that if skilled workers want to keep their jobs in a more sustainable, labour-intensive economy, they should be allowed to do so, even if this means higher prices in some cases. In many European countries there is legislation requiring consultation between unions and management before any labour-displacing technology is introduced. But in the UK, despite the usual flurry of White Papers and Green Papers, nothing has been done. Parliament must assert its right to influence and, where necessary, to control technological developments, at the very least by requiring some kind of 'technological impact statement' for all major advances, and more fundamentally by setting up the equivalent of the US Office of Technology Assessment. Such assessment mustn't just provide a 'value-free' rubber-stamp endorsement of the technocrats of this world, but must be based on full public participation and the liveliest possible debate of the issues and implications involved. Such developments should not be interpreted as romantic protests against progress and science;

there is no suggestion that we have to go back if we are to go forward. As Narindar Singh puts it: 'If it is a protest, it is not against science, but the misuse of it, not against progress but the perversion of it, not against development, but the distortion of it, not against prosperity, but the misconception of it.'[3]

Not the Dual Economy!

When people talk about the economy, they usually mean the *formal* economy, the conventional institutionalized part of the economy, based on full-time taxed employment at a fixed rate. Little attention is paid to the vast amount of *informal* economic activity that goes on in the household and neighbourhood sectors of the economy, some of it paid and some of it not. It should be stressed that 'formal' and 'informal' do not really describe different sectors of the economy, but different ways in which economic activity goes on, with many people participating in both. By and large today's economy looks like this:

    a large-scale manufacturing sector
    a large-scale services sector
    a small-scale local sector
    a household/neighbourhood sector – paid
    a household/neighbourhood sector – unpaid.

Formal economic activity is mostly to be found in the first two, and informal in the last two. The small-scale, local sector combines both, particularly in terms of casual and part-time work, self-employment, and the moonlighting and barter of the 'black' economy.

Industrial societies have always concentrated their attention on the formal economy. It is here that the wealth is 'created' to finance our public services. Over the years, people have been driven increasingly out of the informal economy into the formal: things that people once did for themselves have now been institutionalized or turned into money transactions. Work has come to be seen exclusively in terms of full-time, fixed *jobs*. But now the formal economy and the labour market can no longer provide these jobs or the services that people have come to expect.

Having conditioned people to believe they should have a job, the
labour market can't provide them; nor can it do anything about
the important work in society that is obviously crying out to be
done; nor is it any longer capable of distributing our national
wealth at all fairly.

The formal economy isn't going to collapse; to make such a
claim is just about as daft as supposing that a post-industrial
economy won't have any industry. But it is going to decline. Not
only will we be able to do more with less, in terms of the new
technology, but there will also be less in absolute terms to be
done. In the short term, we shall have to maintain our ability to
compete if we are to earn enough foreign exchange to buy the
primary goods we still have to import, but this sector of the
economy will require continual improvements in productivity,
and will therefore create very few additional jobs. As the numbers
of impoverished and demoralized unemployed grow, we shall
have to face up to choosing between an extension of the socialist
welfare state, and a 'capitalist technocracy' run by the big
corporations in the name of free enterprise. They would have
much in common: centralized economic and political power; an
obsession with technology as a panacea for life's ills; the
dominance of a bureaucratic elite; and the imperatives of dehumanized materialism.

This, I suspect, is what people really mean when they talk
about the 'leisure society'. To me, such a society bears an uncanny
resemblance to the 'dual economies' of the Third World, where
those fortunate enough to be part of a modern, Westernized
money economy lord it over those stuck in the traditional
subsistence economy. The split between work and leisure would
be accentuated, and without work how would people be able to
afford the leisure? Enforced leisure on an inadequate income is a
most unattractive proposition, and pays no attention at all to the
problem of how the social and psychological functions of work
are to be reinterpreted in this brave new technological world of
ours. To call such a society a 'post-industrial society' is quite
absurd; for it is no more than the terminus of industrialism, where
technological determinism finally replaces the political process.

It will not be the leisure of affluence that awaits us, but the
leisure of poverty and political subservience. With such notions of

'duality', ecologists should therefore have no truck. If we are to avoid today's increasing polarities and tensions, any economic strategy must have as one of its *aims* (not as a desirable spin-off) the extension of equity and democracy in our society. This will be possible only if people are helped to liberate themselves from their present dependence on the formal economy, to learn to create *use* value rather than *exchange* value through their work, and to discover how to work in harmony with, rather than against, the interests of the planet.

Balancing the Power

We have already seen how the workings of the market simply can't be counted on to assist in the transition to a sustainable society, even though price rises in commodities and raw materials, combined with the law of diminishing returns, will ensure that people are paying more realistic prices for the natural wealth of the planet. The 'tragedy of the global commons' shows us time after time that the market is no respecter of the Earth's carrying capacity, and its inability to take any kind of long-term view ensures that private, not public, interests are served first. Moreover, it is no respecter of the social and environmental costs inflicted on us today, and so successfully manipulates people into patterns of conspicuous consumption as to ensure that these costs must go on rising.

The Government therefore must intervene, using the full range of sticks and carrots at its disposal, to address the root causes of our current crisis, not the symptoms. Through legislation, direct regulation, changes in the taxation system, subsidies, grants, loans, efficiency standards, the Government has it in its power to effect the sort of transition I am talking about in this chapter. It alone has the ability to resolve the crippling paradox of unmet needs and willing but idle hands. Yet the real fear is that it is so deeply embedded in the pockets of large-scale industry that it will be unable to meet the challenge. Central government, of either right or left, could simply degenerate into the mere agent of the formal economy and its monopolistic, quasi-totalitarian corporate power. Government agencies all too often acquire a stake in maintaining the status quo, and the example of the military-industrial complex provides a worrying indication of the way things may develop.

Since formal economic activity is still the mainstay of prosperity in the UK today, we must strike a balance between protecting the formal economy in the short term, while building up the informal economy for the long term. A massive yet locally controlled investment programme aimed at greater self-reliance would be the first step towards a more sustainable society, covering energy conservation, development of renewable energy sources, housing, urban reclamation, more efficient transport systems, renovation of the sewerage system, forestry, small-scale farming, the clothing and shoe industries, pollution control, waste sorting and recycling, and the repair and expansion of our canal system. It is self-defeating to try to prop up parts of the formal economy that are *not* essential to the future well-being of this country. The lame ducks of today will be the dodos of tomorrow.

There will still be considerable scope for growth within the transitional economy, as we encourage the development of those sectors of the economy needed to secure the future. But the transitional economy must be regenerated *from the bottom up*, under greater local and democratic control, with more emphasis on the creation of socially useful products and the provision of the primary needs of society through small-scale enterprise. Local government should become much more involved in helping to shape the economy of the future, developing an imaginative and flexible framework for the funding of new jobs in small businesses and co-operatives, capable of quick response to rapidly fluctuating local needs. Long-term economic security can be achieved only through local production for local needs; centralized economic power must therefore be devolved and restored to its rightful place – to the community and to the individuals who make up that community. This will require a strategy of positive discrimination in favour of the informal economy so as to diminish the power of big business today, especially the power of the multinationals. We cannot expect small firms and co-operatives to operate successfully in an economy which gives all the advantages to big business. For example, stringent new anti-trust laws should replace existing monopoly legislation – the onus will be on organizations to prove that any merger is demonstrably in the public interest; it will not be enough just to prove that it is not against it.

Reducing Demand

As ecologists, we also lay great stress on the element of *self-reliance*, at both the national and the local level. Since the onset of the Industrial Revolution, world trade has built up increasingly complex patterns of interdependence, leaving many countries in a highly vulnerable position. A lot of unadulterated nonsense is talked about the advantages of free trade, despite the fact that it is neither free nor particularly advantageous. The 'roller-coaster of world trade' rides roughshod over genuine human needs, and in the process of disrupting every local society and ecosystem on the planet causes untold social, cultural and ecological damage. Fortunately, the roller-coaster may itself be grinding to a halt: Third World countries are less prepared to trade on disadvantageous terms, and the transport of bulky goods over long distances will soon become prohibitively expensive. The development of recycling techniques and renewable sources of energy will reduce trade in fossil fuels and raw materials. Despite such developments, it's clear that selective protection of the domestic economy will be needed to establish its sustainable basis, and to encourage this country to become far more self-sufficient than it is at present.

At the local level, families and communities will need to reduce their dependence on outside sources for supplies of food, energy and raw materials. Self-employment should be seen as a major element in new patterns of work, and the home itself, which at the moment has to look beyond itself to supply every need, must increase its economic autonomy. Recreating the domestic economy is *not* turning back the clock; it is a means of ensuring security of supply and protecting oneself against inflation.

The joy of green economics is that there's no trade-off between unemployment and inflation – work is created by the very process of eliminating the root causes of inflation. With economic power restored to its rightful place (to the individual at work in his or her community), the folly of demand stimulation can finally be ended. If you want one simple contrast between green and conventional politics, it is our belief that quantitative demand must be *reduced*, not expanded. With a system of local production for local needs, and with more diverse patterns of fulfilment and reward, many of the causes of inflation would be removed: the ability of domestic and international monopolies to pass on rising costs to the consumer; the continual pressure for higher wages brought about by economic and social inequality; central government's over-enthusiastic reflating of the economy, in the forlorn hope of achieving full employment. And there are many other ecological factors that will assist the fall in demand: population stabilization, a shift away from consumerism, the move to renewable energy sources, the expansion of the self-help economy, an end to built-in obsolescence, massive reductions in defence spending, and an emphasis on conservation, reuse and recycling. As we reduce the complexity and 'external costs' of our industrial society, referred to at length in chapter 6, so again inflation will fall. And as we develop an economy as self-reliant as possible in primary goods – which are fast becoming the hard currency of the developed world – our vulnerability to world inflationary pressures will be reduced.

I said at the start that green economics is not in the business of dressing up panaceas as a means of proving our political virility. Yet when you look at things quite logically, without the distortions of the grey-tinted spectacles of industrialism, it's astonishing just how many answers there are to be found within it! But dealing with unemployment and inflation requires a great deal more than economic wizardry. It was Pierre Trudeau in 1977 who said: 'Inflation has not found its Keynes. I personally think the Keynes of inflation will not be an economist, but instead a political, philosophical or moral leader, inspiring people to do without the excess consumption so prominent in developed countries.'[4]

Minimum Security

Despite a residual confidence in the ability of people to think logically about these things, I can't help but acknowledge that the transitional period is going to be pretty stormy. Many people feel increasingly insecure, and one of the main reasons for this is that the labour market no longer works very well as a means of distributing wealth. At such a time I can see no option but to guarantee basic economic security through the introduction of some sort of Social Wage or Minimum Income Scheme. Various proposals for such a scheme have been around for a long time; the simplest would be to replace all social security benefits and all tax allowances with a single, automatic payment to everyone which can be taken in cash or used as a credit against tax liability. Such a payment would be made without qualification, at levels varying according to age and circumstances, so as to guarantee minimum subsistence.

In the welfare state today, basic needs are guaranteed for all *as a last resort*; the Minimum Income Scheme would mean that basic needs were guaranteed *unconditionally*, but that there would be competition for anything else over and above that level of subsistence. Any claimants today who try to stand on their own two feet very quickly have the welfare rug pulled away from under them, with benefits given on the condition that they do *not* work. A Minimum Income Scheme would remove work as an obligation, but would reinstate the work incentive taken away by the welfare state. In so doing it would remove the poverty trap once and for all, by combining a sense of personal security with the incentive for initiative and effort. The effects of such a scheme on unemployment could be dramatic: hours of work and rates of pay would become more flexible and employers could take on more employees for the same wage bill. The self-employed would be encouraged to try new ideas. Above all, instead of the dreadful distinction between being employed and being unemployed, there would be a continuum from those content with basic subsistence to those willing to work long, arduous hours in return for certain material benefits.

It goes without saying that such a scheme would cost a lot of money and would therefore require a radical shift in taxation. But the dilemma of achieving a genuine redistribution of wealth in this country must be faced sooner or later. Growth has become a substitute for equality of income, making tolerable large differentials in wealth; in an economy that's *not* growing in the same way, redistribution is a precondition for any transition to a stable society. (In the long term, I personally believe it will be necessary to think in terms of maximum income as well as minimum income schemes.) It is also important to realize that without a Minimum Income Scheme, the poverty trap will remain with us and will become more serious as the levels of unemployment in the formal economy rise.

With a flourishing informal economy, and guaranteed basic security, many people who would like to be doing other things for a part of their working life would not be prevented from doing so. Work-sharing is already a vital necessity; only in such circumstances can it become a reality. We face a choice between an economy in which 20–30 per cent of the population work between forty and fifty hours a week, are highly paid and highly taxed in order to maintain the rest and an economy in which those who are able and willing to work have the right to such work, but for fewer hours, thus spreading both work and its benefits more equitably through society. There are many ways in which this could be achieved: by shortening the working day, the week, the year or even the working lifetime. One thing that seems crazy to many people is the fact that so much overtime is still worked in this country. In the manufacturing sector about one-third of the labour force regularly works eight hours' overtime a week – if all those hours could be parcelled into full-time jobs, this would immediately absorb almost all the registered unemployed in this sector. In other areas, job-sharing and part-time working should be widely encouraged, and special schemes could be devised for parents with young children to enable them to work part-time within the community.

But you can't get something for nothing in a shrinking economy, and such a strategy would have to be paid for. Without a $12\frac{1}{2}$ per cent increase in productivity, a thirty-five-hour week at a forty-hour week wage is inevitably inflationary. Work-sharing can be both self-financing *and* employment-creating only if we acknowledge that it will inevitably mean a reduction in that proportion of income derived from work within the formal economy. That will be acceptable to people only if they see not only that are there financial compensations in the informal sector, but also that they would gain in terms of less tangible benefits such as job satisfaction, genuine quality of life, etc. Any attempt to fudge this issue by concealing the necessary redistribution of wealth through inflationary wage settlements would be totally self-defeating.

The Agents of Change

Rising unemployment and a growing awareness of the dis-economies of scale have obliged politicians of all parties to reappraise the role of the small businesses. They have become the centre of political attention and flattery, but typically little has actually been done to provide them with real support rather than empty rhetoric. David Birch of the Massachusetts Institute of Technology, commenting on the fact that small firms generated two-thirds of the new jobs created in the USA between 1969 and 1980, added this insight: 'The job-generating firm tends to be small, it tends to be dynamic (or unstable, depending on your viewpoint), the kind of firm banks feel very uncomfortable about... in short, the firms that can and do generate the most jobs are the ones that are the most difficult to reach through conventional policy initiatives.' In the kind of long-term economy that we envisage, small businesses would not just be a useful adjunct to the world of corporate big business: they would be the mainstay of all economic activity.

There are 101 things that should be done by government, and 101 things that could be done by small businesses themselves. I was particularly inspired by reading about the Briarpatch Network in the USA, an informal association of people and businesses who work together on the basis of shared values. They see business as a way of serving others, not primarily as a means of making the largest possible profit, and are concerned that all book-keeping records should be open and available to the customer. The network provides advice, support, and a way for people in business to come together convivially. The survival rate of businesses in the network is extraordinarily high, and new networks are being set up all the time.

There is one particular form of the small business that is especially important in the eyes of ecologists, and that is the co-operative. A co-operative is much more likely to be sensitive to the needs of the community in which its members live. The profit motive is linked to a broader collective concern: concern on the one hand that the working members are adequately cared for, and on the other that the co-operative is playing a constructive part in the wider community. Regrettably, the co-operative movement has had a tough time in Britain, in sharp distinction to many other European countries. The labour movement itself must take a considerable share of the blame for this, for in its obsession with nationalization it has succeeded in making the whole area of 'common ownership' suspect and unattractive. For many, common ownership is now synonymous with nationalization – but nationalization has never even pretended to give employees increased control of or influence over their work, and the management of the nationalized industries is identical to that of their autocratic counterparts in big business.

But things *are* looking up: a growing number of industrial co-operatives are being established, and their survival rate compares well with that of small businesses generally. There has been legislation in the form of the Industrial Common Ownership, Inner Urban Areas, and Co-operative Development Agency Acts; a number of local and national co-operative development bodies have been created, and many individuals and organizations now recognize that the encouragement of common ownership schemes and co-operatives is vital. But if the ethos of the co-operative movement is significantly to influence the general industrial climate, then a policy of positive discrimination is urgently needed.

The example of Mondragon has had a considerable impact in this country. Set in the midst of the Basque region in north-west Spain, this hugely successful initiative now involves about 20,000 people operating in about 200 trading co-operatives and several housing and agricultural co-operatives. The ownership and control of each enterprise is confined to the people working within it, but the collective as a whole has built up certain shared services, including its own social security and health organizations, and a most progressive technical college. Between them they have managed to weather the recession better than many of their fellow Europeans in or out of conventional employment.

The key to the success of the Mondragon Common Ownership movement has been the local co-operative bank, which now has ninety-one branches throughout the region. It seems obvious that in order to finance a locally based economy, we shall have to establish a locally based network of Community Savings Banks. Initial finance could be provided in part by a 'windfall' tax on the totally unjustifiable profits of the big banks, and in part through a temporary scheme whereby, for each deposit held by the Community Savings Banks, the Government would add a certain percentage (probably derived from North Sea oil revenues), thus assisting directly in the creation of local employment. Once established, these banks would operate in a more or less conventional way, but on a *non-profit-making basis*. They would offer conventional services to attract deposits, and would be permitted to vary their interest rates both to savers and borrowers in order to attract and use funds. They would not, however, issue loans for consumption, but only for the purpose of capital investment and the creation of viable local enterprises. The immediate job-creating potential of such banks is far greater than central state expenditure on huge, wasteful projects. The initiative for regenerating the economy would be coming from local people with local knowledge, and the whole community would become involved in the creation of real, long-term wealth, rather than the spurious 'wealth' of advertisement-induced mass consumption.

Simultaneously, we believe that a whole series of Regional Enterprise Boards should be set up to encourage socially useful production at the local level, and facilitate a revitalized co-operative structure to arise out of the decay of existing industries. In certain cases, funding may be needed for projects beyond the scope of purely local investment; sizeable sums of public money committed *now* would reduce the likelihood of chaos in the future. In such a context a comprehensive training strategy could seek both to equip school-leavers with marketable skills, and to provide older workers with the opportunities to acquire new skills. Retraining needs not only to be greatly increased, but also designed to encourage a more 'interdisciplinary' approach, so as to keep pace with the high rate of innovation and to find new ways of using skills made redundant by technological advance. It will also be necessary to promote those skills which will be in demand in the more self-reliant, 'do-it-yourself' atmosphere of the transitional economy, encouraging 'skills exchange networks', so that those who want to teach and those who want to learn are put in touch with each other.

The Old and the New

Green economics is an attempt to blend the old wisdom of the Earth with the dynamic potential of new technologies. Hence the encouragement of new industries (for in the transition to the sustainable society these will emerge as rapidly as the old ones fade), and the emphasis on new skills and new ways of doing business. But ecologists combine this new-age sophistication with the realization that the finite resources of the planet will simply not stretch to sustaining endless economic activity carried out purely as a means of providing work, increasing profits or raising the levels of GNP. Hence the parallel emphasis on the primary sector of the economy, which in today's glossy, abstracted world may seem somewhat unfashionable. Most economists and politicians conveniently manage to ignore the 'primacy' of our agriculture, raw materials, mines, forestry, energy resources and fisheries, but this will inevitably change as we are forced to base our economy on the pursuit of sustainability rather than growth.

Throughout the rest of this part of the book the implications of such changes in attitude should become clear. The mind boggles at the amount of work that so urgently needs to be done, not only to enhance the self-reliance and sustainability of our economy, but also in areas where human skills and abilities are at a premium, areas which will remain labour-intensive despite the new technology. New ventures and projects are already springing up all over the place, particularly among the unemployed; this is one instance where the practice often precedes the theory.

It should also become clear how a more decentralized, sustainable society, in which the overheads and the running costs of today's astonishingly wasteful, unhealthy and alienating way of life have been all but eliminated, would be immeasurably more 'economic' in the true sense of the word. None the less, considerable sums of money will certainly be needed to finance the transition. Unless the money is spent *now*, ensuring a peaceful and socially acceptable transformation, it is absolutely certain that it will not be available in the future. It is therefore imperative to use our diminishing capital wealth, and the even more rapidly diminishing returns from a growth-oriented economy, to effect the transformation before it is too late. Of greatest importance, in this respect, are the revenues from North Sea oil. The use we make of this 'windfall' is likely to be the determining factor in successfully bridging the gap between where we are now and where we will need to be by the year 2000.

The dilemma is simply stated: for every year we delay making a move in the right direction, the consequences become proportionately more serious. Developed and Third World countries alike must seek to map out a different course for themselves, renouncing the maximization of production and consumption based on non-renewable resources, moving towards a sustainable society based on renewables and the elimination of waste. The challenge is to meet the inner demands of basic human needs without violating the outer limits of the planet's wealth. The old system is bankrupt, and it is only the wisdom of ecology that will show us how to create a new economic order.

I I
Green Peace

Ten Disarming Reasons

Ever since I first heard the story of Ajax in Nicholas Humphrey's
1981 Bronowski Memorial Lecture, it's stuck in my mind as the
grimmest of allegories:

> When I was a child we had an old pet tortoise called Ajax. One
> autumn, Ajax, looking for a winter home, crawled unnoticed into
> the pile of wood and bracken my father was making for Guy
> Fawkes Day. As days passed and more and more pieces of tinder
> were added to the pile, Ajax must have felt more and more secure;
> every day he was getting greater and greater protection from the
> frost and rain. On 5 November, bonfire and tortoise were reduced
> to ashes. Are there some of us who still believe that the piling up of
> weapon upon weapon adds to our securtiy – that the dangers are
> nothing compared to the assurance they provide?

How can it possibly be that so many people today act like
Ajax? Why are they still persuaded that the best way of ensuring
security is to go on spending more and more on arms? How can
politicians, of any party, allow their judgement to be so
terminally impaired that most of them are still out there laying on
the firewood? Even when all the dangers are acknowledged,
people are still motivated by a fear that outweighs all their reason
and awareness: the fear that the enemy is intent on outdoing us.
That makes our weapons 'good', their weapons 'bad'; our
strategy 'defensive', theirs 'aggressive'. Jim Garrison quotes the
British physicist P. M. S. Blackett: 'Once a nation bases its

security on an absolute weapon, such as the atom bomb, it becomes psychologically necessary to believe in an absolute enemy.'[1] And the Cold War warriors, with their snarls of paranoid hatred, ensure that we remain psychologically armed to the teeth.

I don't happen to believe that there is any great threat from the Russians, as I tried to demonstrate in chapter 5. The whole notion of their launching any kind of attack on Western Europe is to me utterly incredible. I realize that this is not the case for many people, and that the arguments against our current defence strategy have to be justified in the light of their continuing anxiety. But I would like to suggest that however sincerely one may fear the Russian threat, the following ten reasons are sufficient in themselves to explain why we must rid this country of nuclear weapons just as fast as possible.

1. The Credibility of the UK Deterrent

A nuclear deterrent is full of paradoxes: to prevent the holocaust, we must be fully prepared to cause it; to be effective, it is dependent on the rationality of both sides wishing to avoid the use of such weapons, and on each appearing to the other side to be irrational enough to carry out the threat to use them! Strategic reasoning has been replaced by pretence and bluff – and if the bluff fails, there is no second chance. As George Kennan, former US Ambassador to the USSR, put it in his acceptance speech for the Albert Einstein Peace Prize in 1981, 'To my mind the nuclear bomb is the most useless weapon ever invented. It can be deployed to no rational purpose. It is not even an effective defence against itself.' If we ourselves ever used our nuclear weapons against a superpower, we would be courting certain and total destruction. To think of commiting suicide in such a fashion is quite incredible; but for deterrence to be effective, the threat to use them has to be credible.

There are those who say that to get rid of nuclear weapons would be to disturb the delicate stability that they ensure. The evidence for this form of intellectual blackmail is not convincing. The burden of proof is surely on those who are threatening genocide, who are preparing to wage nuclear war, for only the *complete certainty* that the threat of nuclear war will prevent the outbreak of nuclear war can possibly justify their preparations. Such certainty is not possible. And there are those who quote the lessons of World War II, when Hitler took advantage of weakness and appeasement on the part of the Allies to devastate Europe, while they choose to forget the lessons of World War I, and the contribution that the huge build-up of arms made to the start of that war. History offers no instance of intensive preparation for war that has not ended in war.

2. The Moral Case

Up until the invention of nuclear weapons, the moral justification for fighting a 'just war' was clear: one had to discriminate between combatants and non-combatants, and the effects of going to war had to be less destructive than the disaster it was intended to prevent. It is utterly impossible to maintain that nuclear war conforms to these principles.

Hence the muddle within the Church of England over its own report, *The Church and the Bomb*. Despite the strong case made for unilateral disarmament initiatives in the report, the 1983 Synod voted for a compromise proposal demanding a declaration of 'no first use' (thus undermining the whole NATO strategy which depends on the readiness to use nuclear weapons first), while confirming that NATO's policy of deterrence was still justified. In this way the Church of England declared its support for the use of nuclear weapons, for deterrence can be effective only if the threat to use them is serious. What residual claim the Church of England still had to moral authority disappeared in that one vote, causing agony to hundreds of thousands of Christians. And how Graham Leonard, the Bishop of London, can reconcile his enthusiastic defence of deterrence with the Archbishop of Canterbury's contention that 'loving our enemies is the only strategy the Church is empowered to pursue' defies my understanding of Christianity. To say that it is permissible to threaten nuclear war but not to wage it is an appalling evasion unworthy of any religious person. The moral cost of upholding the theory of deterrence is that we have all of us underwritten the possibility of massive slaughter. To be a target is bad enough, but to be involved in the targeting of others is much worse. As Jonathan Schell points out, 'In accepting the dual role of victim *and* potential mass-murderer, we convey the steady message that life is not only *not* sacred, but utterly worthless.'[2]

3. Stewards of the Earth

Jonathan Schell's book, *The Fate of the Earth*, spells out exactly how devastating the impact of nuclear war would be on the whole biosphere, not just on humans. We would literally be destroying the future. The 1983 Washington Conference on the Long-Term, World-Wide Biological Consequences of Nuclear War more than confirmed Schell's impressions. This vitally important conference, bringing together the work of hundreds of eminent scientists in both the USA and the USSR, unanimously agreed on several conclusions. In the event of a large-scale nuclear war a pall of darkness caused by dust and the smoke from fires would cover the northern hemisphere, and spread rapidly into the southern hemisphere. Sunlight would be drastically reduced, halving plant growth, and causing a 'nuclear winter' through dramatic falls in temperature to sub-freezing levels for several months. Exposure to radioactive fallout would be much worse than previously anticipated, on account of the pall of smoke and dust, and would last for many weeks. The ozone layer would be irreversibly damaged, increasing the exposure to ultra-violet light, causing blindness, widespread skin cancer and grave damage to the immune system of humans and all mammals.

Even a relatively small nuclear exchange would cause severe after-effects. Even a successful first-strike, eliminating the other side's weapons, would be committing suicide, since the effects of the nuclear winter would be just as severe for the 'winner' as the loser. No part of the planet would be safe from these effects. If we are prepared to accept our role as stewards of the planet, of all other species, and of the future itself, then it must be stated once and for all that there is no single conceivable situation which could justify the use of nuclear weapons. *Nothing*, but *nothing*, would excuse the irretrievable damage done to the Earth itself.

4. Multilateralism or Unilateralism?

During the 1960s, the Pentagon reckoned that 400 megatons would be sufficient to destroy the Soviet Union's industry and most of its population. Today, despite all the talking, the combined destructive capacity of the superpowers' strategic weapons is more than 15,000 megatons. Constructive emphasis on unilateral initiatives is far more likely to lead to agreement than the current policy of 'negotiating from strength', which basically operates as a cover for the continuing arms race. The lack of progress in multilateral negotiations is often used as a pretext for acquiring new weapons. As Martin Ryle says in his *Politics of Nuclear Disarmament*:

> It goes without saying that genuine multilateral disarmament would be preferable in every way. Indeed, multilateral disarmament would be the goal of such unilateral initiatives. Meanwhile, the choice – the initial practical choice – is not between unilateral and multilateral disarmament; it is between unilateral disarmament and no disarmament at all. Every nuclear state since Nagasaki (in so far as they have even acknowledged the need for disarmament at all) has claimed to favour a 'multilateral' approach, and has 'pursued' nuclear disarmament within that framework. Those who argue that this kind of diplomacy is suddenly going to produce real progress ignore the blatant evidence of history, and must be convicted either of naivety or insincerity.[3]

Since the start of the nuclear arms race, not one single nuclear warhead has ever been negotiated away.

5. Proliferation

I have already discussed the extreme dangers of proliferation in chapter 5. The responsibility of Britain for the spread of nuclear weapons, through both our own example and our nuclear energy policy, can only be viewed with the deepest shame.

6. War by Accident?

As we improve the speed, range and accuracy of our missiles, just how hot will the hot line prove to be? The Pershing II takes just eight minutes from launch to target; the USSR cannot wait to check things out – it will have to 'launch on warning' if it is to ensure that its missiles are not caught in their silos. Whether the initial launch was unauthorized or an accident will by then be quite academic. False alerts and computer faults are common- place; Sod's Law tells us that one day something is bound to go badly wrong. In the *Disarmer's Handbook*, Andrew Wilson details thirty-five recorded accidents in the West, due to human error or systems malfunction. Officials often express alarm at the extra- ordinarily high incidence of psychological disorders, drug abuse and alcoholism at Strategic Air Command's missile bases – little wonder, given the stress and the schizophrenic reaction to the thought of carrying out mass murder.

7. The Death of Deterrence

Those who still support Britain's dependence on nuclear weapons usually do so on the basis of their *deterrent* value. Yet it has *never* been the case that the sole purpose of NATO's nuclear weapons is to prevent nuclear war: a separate purpose has always been to prevent a Soviet military victory. NATO's strategy of 'flexible response' means that we would be prepared to use nuclear weapons first as an integral part of our response to a conventional attack. The USA and Britain have persistently refused to commit themselves to any 'no first use' declaration. Despite having no evidence whatsoever, the USA has always justified its arms build- up by claiming that the Soviet Union might at some stage contemplate a first strike. On the contrary, *all* the evidence indicates not only that the Soviet Union has *never* contemplated this, but that the whole thrust of US doctrine and scientific research has been to enable it to mount a first strike itself.

In 1980, Presidential Directive 59 enunciated the twin notions of 'limited nuclear war' and 'counterforce' – meaning an attack on selected military rather than civilian targets. There's no point in firing at the other side's weapons – the basis of counterforce – unless they're still in their silos. That means striking first.

In the massive US defence budget for 1985, the largest increases have been earmarked for nuclear war-fighting programmes and the development of weapons with 'hard-target kill capability' (i.e. counterforce). The strategy of MAD (Mutually Assured Destruction) is out; the old military logic of attack being the best form of defence is back in. Nuclear weapons are no longer designed to *deter* the Soviet Union; they are being designed to *defeat* it.

8. 'Limited' Nuclear War?

The limitation in this instance is the boundaries of Europe; the idea behind it is that within this theatre, the superpowers could slog it out without actually involving direct attacks on their own territories. There are some who believe that this is a credible possibility. US Admiral La Roque: 'We fought World War I in Europe, we fought World War II in Europe, and if you dummies let us, we'll fight World War III in Europe.' There are many more who believe it's a ridiculous notion, including the Russians, NATO Supreme Commander General Rogers ('The use of theatre nuclear weapons would in fact escalate to the strategic level, and very quickly'), and most reasonable people. *Either way*, mind you, Europe is totally destroyed.

As I've already outlined, the notion of the 'limited war' has now been taken one horrifying stage further to include the notion of the 'winnable nuclear war'. This is official US Government policy, assiduously promoted by the closest advisers of President Reagan. The 'Holocaust Lobby' still believes that 'nuclear war is unlikely to be an essentially terminal event.' It has just been revealed that the Pentagon is already planning the development and deployment of reserve nuclear weapons for the eventuality of World War IV.

9. Civil Defence

Part of the strategy to make nuclear war 'less terminal' is the active preparation of civil defence facilities. The USA has set aside $4 billion this year as part of the new *offensive* nuclear war strategy. An official US civil defence manual states: 'Victory in a nuclear war will belong to the country that recovers first.'

A statement of such colossal idiocy would come as no surprise to the government officials in Britain who drafted the comic booklet *Protect and Survive*. The Government knows it's lying about the consequences of a nuclear attack (as demonstrated by its decision to cancel the civil defence exercise, 'Operation Hard Rock'), but part of a nuclear war-winning strategy is that people must be persuaded that they *can* survive it. The Government's 'expert scientific advice' has now been totally refuted by the British Medical Association. A credible civil defence programme is impossible (other than as a way of maintaining law and order through the imposition of authoritarian controls); even if the number of available shelters were increased by hundreds of thousands, what would people emerge into after a nuclear war? With no food, industry, sanitation, medical facilities, fresh water, communication, transport or energy, the living wouldn't actually have much time to envy the dead.

10. The New Weapons

Nothing more clearly demonstrates that technology now controls strategy than the new counterforce weapons of Cruise and Trident. Cruise missiles are said to be necessary to balance the Russian SS-20s. The SS-20s are a modern replacement for the SS-4s and SS-5s, first introduced in the 1960s in response to the *unilateral* decision of the US Government to base tactical nuclear weapons in Europe. The SS-4s and SS-5s were *always* capable of reaching targets in Western Europe, though we are now told that we need Cruise specifically to match this 'new' capability of the SS-20s. NATO knew about the SS-20s for years without expressing *any* concern, and their apparent 'threat' was not raised until the US Defense Department was on the point of signing the contracts with Boeing for the production of Cruise!

Cruise missiles basically make arms control impossible because of their size and flexibility. They are also particularly worrying to the Soviet Union because of their accuracy (to within 30 metres) which threatens the 'command, control and communications' facilities on which the USSR depends. The Americans have just allocated $18 billion to protect their own communications facilities as an essential part of their new war-fighting strategy. And though we in Britain are particularly concerned about ground-launched Cruise missiles, of even greater concern are the sea-launched missiles which the USA plans to deploy in their thousands, on every conceivable kind of naval craft.

In the UK our primary concern should now be Trident. We are to get at least four of these 'insurance policies', each one capable of unleashing 224 warheads, with each warhead twenty times as powerful as the Hiroshima bomb, and accurate to within 90 metres. They are first-strike weapons, whereas Polaris was quite clearly retaliatory. This is unilateralism at its very worst: unilaterally upping our 'defence' capabilities through the hugely expensive purchase of an inappropriate, counterforce weapons system. Field Marshal Lord Carver has summed up the opinions of most thinking people: 'It would be suicidal for us to threaten to use Trident against Russia. So what the bloody hell is it for?'[4]

Life or Death?

I simply don't understand how anyone, faced with all that, can still countenance Britain's continued dependence on nuclear weapons. Unilaterally to get rid of ours seems the *only* option, and then by strenuous and genuine multilateral negotiation to start reducing the stockpiles of the superpowers.

However much we may abhor the Soviet Union, however much we may still fear it and its appallingly oppressive system, we have to learn to live in peaceful coexistence. We either learn to live together, or we die together. Raising the level of hostilities will only make things worse in a country that already feels itself to be surrounded by enemies on all sides. Moreover, arms negotiations cannot get anywhere until we arrive at a consistent policy for friendly relations with the USSR. The old saying 'If you want peace, prepare for war' is of little use in today's world with the weapons we now have.

For it is not only nuclear weapons we must think of. We are threatened with the equally hideous consequences of chemical and biological weapons. The USA has recently approved massive funding for the production of a new generation of 'binary' chemical weapons, with the two components being stored separately until use. It claims that the USSR has a huge lead in such weapons, but there is absolutely no evidence of this. Talks about a Chemical Weapons Treaty were until recently bogged down in the thorny problem of how to organize on-site inspection and verification, but the gap between the two sides now seems to be narrowing. Both must have been chastened by Iraq's use of chemical weapons in the war against Iran, and the ease with which such a country was able to develop such appalling weapons in a plant ostensibly build to produce pesticides.

All defence policies are a risk, but some risks are infinitely greater than others. The only question that matters is which course is most likely to promote peace. At the moment we are joined in a mutual suicide pact, and the first risk we in Britain must take is to accept the evidence that there is a strong desire for peace, both here and in the USSR, and unilaterally get rid of our nuclear weapons. It's impossible to see what possible advantage there could be for the USSR in attacking or threatening to attack the West. But if such a thing did happen, then we would indeed have to capitulate and endure occupation.

To which some, even now, might say, 'Better dead than red.' I find this the most astonishing proposition, and when I hear it, I know that reason has fled. The defeat of the state does not mean the surrender of the people; how can one opt for extinction rather than hope, however small that hope may be? Would the people of Poland prefer to be buried in mass graves as victims of a nuclear war, rather than fighting against the oppression and exploitation inflicted on them? There's something cowardly and ignoble about such an attitude, for one does nothing to protect human freedom by committing suicide.

Non-Alignment

Once that all-important threshold is crossed, from a nuclear to a non-nuclear defence strategy, the rest more or less falls into place – on the basis of logic.

First, we would have to withdraw from NATO and get rid of all the American bases in this country – more than a hundred of them. Those who argue that we should get rid of our own weapons, and yet remain a member of an alliance whose strategy is wholly dominated by nuclear weapons, are either kidding themselves or deliberately trying to kid others. The whole purpose of nuclear disarmament is to avoid getting wiped out in any war by not constituting a threat to any potential aggressor; our objections must be not just to the British bomb, but to the whole concept of nuclear deterrence, and our aim must be to minimize the risks of war in the European theatre. *Permanently* stable nuclear deterrence is quite impossible.

NATO is more likely to be the cause of war than the agent of peace. At a time of considerable political stability in Europe, it is quite erroneous to suppose that it is the presence of NATO that has prevented war. By its wholesale commitment to the Cold War, it has promoted an exaggerated fear of the Soviet Union for reasons of its own, and it has been instrumental in the largest build-up of arms the world has ever seen. In its excellent report *Defence Without the Bomb*, the Alternative Defence Commission recommended staying in NATO *conditionally*, so that we could negotiate from within to move NATO towards a non-nuclear strategy through a 'no first use' policy, the withdrawal of all battlefield and tactical weapons, and the decoupling of NATO from US nuclear strategy. Given the dominance of the USA in NATO, such a policy is really a complete non-starter, an illogical hankering after the old notions of security even as one moves towards the new. Europe must decouple itself entirely from the influence of the USA; we do not want to be part of its nuclear umbrella, and we want to be free of its whole approach to foreign policy. Its bases in this country have turned us into an advance aircraft carrier, and have made us the number one nuclear target in Europe. They must all be phased out, regardless of the political and economic pressure that the USA will no doubt bring to bear.

Our withdrawal from NATO might well bring about its break-up, but that will assuredly happen anyway as other countries come to reject dependence on nuclear weapons. Some people believe that it would be important to set up some sort of alternative structure (a European Defence Association) to safe-guard our links with and commitments to Europe; but there seems to be little point breaking up NATO merely to set up another integrated, multinational force with a unified command structure, even if it were to be based on conventional weapons. Such an association would still perpetuate Cold War tensions, and the attempt to create another great power to rival the USSR and the USA would be prohibitively expensive and unlikely to promote peace.

It might, on the other hand, be both practical and sensible to set up a much looser kind of association, relying primarily on the national defence forces of each member state, but guaranteeing certain kinds of support and solidarity should any attack occur. I can see no incompatibility between such an association and the policy of non-alignment which I believe we should adopt, for countries would be agreeing to support each other in the event of aggression from *any* source, East or West. It would still be possible to develop a totally independent defence policy within such a framework, and at the same time to act as a positive force for peace, offering active mediation in arms negotiations and sudden conflicts, and setting up the sort of links that will allow the nations of both Eastern and Western Europe to coexist peacefully.

Non-Nuclear Defence

So there we are, non-nuclear and non-aligned. Do we then go the whole hog, get rid of all our conventional forces and rely entirely on non-violent civil resistance? After all, if there's *no* threat from the Soviet Union, why bother with *any* defence? And even if there is a threat, why bother? We could neither withstand the threat of a nuclear attack nor win a conventional war. So wouldn't it be logical to dispense with all conventional weaponry?

I think *not* – not *yet*, at least. Even though I totally reject the stereotypes of the Cold War, I can still see that in a dangerous and volatile world the mere existence of the superpowers poses a *potential* threat. I can envisage situations in which either of them might see fit forcibly to seize territory or certain facilities for strategic reasons. Britain's position in the North Atlantic gives us considerable strategic importance, and in the desire to pre-empt possible moves by one superpower, the other might well be tempted to move against a state that was undefended, but not think it worthwhile to engage in a prolonged military campaign or to suffer the international humiliation of having to threaten to use nuclear weapons. A strategy of civil resistance would be unable to prevent this, nor would it be able to do anything about a possible blockade or the seizure of off-shore oil rigs. And who is to say, in a world dominated by resource shortages, that present allies will remain allies for ever?

Moreover, civil resistance is currently not a realistic option in the UK. Most people are completely ignorant about it, and the vast majority still believes that war is justified in certain exceptional circumstances. It strikes me that the *only* way to persuade people to reject nuclear weapons is by ensuring that for the time being we retain a viable conventional defence capability. If we push too far too quickly, it will be counterproductive, and the peace movement would probably disintegrate. This is more than a debate about tactics. I very much respect the views of those who reject the use of *all* weapons, and though I myself am not a pacifist, and can just about imagine occasions when I would be prepared to fight for the ideals I believe in, I am painfully conscious of the contradiction between this and my commitment to the principles of non-violence. I am also conscious of a gradual change going on in my own attitudes as I feel myself moving nearer a pacifist position. I suspect that this sort of gradual change is going on in a lot of people, and would therefore suggest that the prime responsibility for greens in the peace movement is to promote this sort of change by arguing the case for non-violent civil resistance while temporarily acknowledging that conventional defence policies have an important part to play in the transition to a more peaceful world.

From a green point of view, such policies must pose the minimum threat to others, must allow for a reduction in military spending and must promote further disarmament initiatives. At the same time they must raise the political and military costs to any aggressor in the defence of our legitimate interests. Hence our emphasis on *defensive deterrence*, which would completely rule out the possibility of our being able to mount any significant offensive operation ourselves. Defensive war enjoys many advantages over offence and recent developments in precision-guided weapons have considerably reinforced this traditional advantage.

A conventional, frontier-based approach, maximizing the advantages of Britain's being an island, would rely on strong coastal and anti-aircraft defences, fighter aircraft, a navy made up mainly of submarines and medium-sized craft, and a highly mobile regular army backed up by an expanded territorial force.

It is vitally important to limit the costs of such a strategy. It is not just a question of reallocating the Trident billions or money saved from withdrawing the British Army of the Rhine: that money is needed elsewhere. We may therefore have to consider putting more emphasis on 'in-depth territorial defence', relying on manpower and cheaper weaponry. But such decentralized defence, with rearguard actions up and down the country, is not particularly suitable for the terrain of the UK, and would involve much greater suffering for the civilian population. It might also require conscription or compulsory part-time military training, to which I'm strongly opposed as an unacceptable violation of personal freedom, and any widespread militarization of society tends to undermine moves to promote methods of non-violent, civil resistance. For the same reasons, I would be rather unhappy about the use of widespread, co-ordinated guerrilla tactics as a fall-back strategy against occupation.

Despite the fact that one still encounters general resistance to the idea that defence and security can be thought of in *non-military* terms, there are good reasons for moving as rapidly as possible towards a strategy of civil resistance – not only as a fall-back strategy in the event of occupation through nuclear blackmail or conventional defeat, but as the central component of our defence. Any modern war is horribly destructive, and the only really effective way of reducing arms spending is to get rid of the arms. As I said earlier, the defeat of the state does not mean the surrender of its people. The aims of civil resistance are to make the country ungovernable for any occupying power, to deny it the economic benefits it would hope to gain, to sow dissent and disaffection among its troops and officials, to maintain the morale of one's own people whilst undermining that of the occupying power, and to encourage international sanctions against that power. Such resistance would be in the form of strikes, boycotts, go-slows, civil disobedience, demonstrations and mass non-co-operation. It might also include selective sabotage of facilities, industrial plant, electronic machinery and communications. There are many examples where such resistance has been extremely effective, even against very repressive regimes.

To my mind, the most compelling argument in favour of civil resistance is its *deterrent* value, either as a back-up to military defence or on its own. The problem for any occupying power trying to impose its will would be staggering; in such circumstances, for the Soviet Union to invade Western Europe would be like taking on the problems of Warsaw and Kabul a hundred times over! This makes civil resistance an extremely hard-headed, logical defence strategy – but *only* if there has been thorough preparation in advance. A Civil Resistance Commission should be set up now to prepare training manuals and develop contingency plans as an essential preliminary to shifting from conventional to civil resistance. The difficulties involved should never be underestimated, for defence of this sort will require a similar commitment and courage; the need for people to die for their country would not disappear merely because we had done away with our weapons.

If prepared for on a collective basis throughout Europe, as an integral part of any informal defence association, such an approach could be particularly effective. It would offer the strongest possible deterrent: the total ungovernability of Britain. It would signal with indisputable clarity that our intention was to promote peace, not pursue war. And it would eliminate our threat to any other country, while freeing vital resources for more useful purposes.

Securing the Future

The greens have at last started to play a more important part in the UK's peace movement, some through their involvement in CND and some through other organizations. We are concerned that the over-identification of the peace movement with left-wing politics has so obviously put an upper limit on the potential effectiveness of the movement, and believe it's crucial that we begin to build a broader base. It is therefore not the time to be worrying about purity of policy but, starting with support for a mutual freeze on the testing, development, production and deployment of *all* nuclear weapons, we must begin to move step by step towards common security rather than mutual annihilation.

But we have to ensure that these steps are actually on the road to *peace*. There's a great deal more to lasting security than either unilateral nuclear disarmament or a viable conventional defence strategy. Such policies must be set within a social, political and spiritual framework which allows for the possibility of peace becoming a lasting reality. The broader green position on this point is uncompromising: if we continue to live the way we do now, lasting peace is simply not possible, with or without unilateral nuclear disarmament. The increasingly stubborn maintenance of today's status quo ensures a high level of continuing violence in international affairs. It is, therefore, the underlying *causes* of conflict that we should also be campaigning to eliminate.

At the moment, people have to choose between fatalistically going on as we do now with the bomb, or fatalistically going on as we do now without it. That's really not much of a choice, and may account for the lack of realism that sometimes seems to characterize the peace movement. Are we just going to put a Bill through Parliament and, hey presto!, no more nukes, and smiling Americans patting us on the back as they cart their missiles back across the Atlantic? The bomb cannot be disinvented; the violence that is inherent in our materialist, industrial culture cannot be eliminated at the stroke of a pen. Nuclear weapons must be seen as a symbol of our failure to build peaceful relations and to use our technological prowess more wisely. And, as we've already seen, the arms race is merely the most deadly extension of a world already at war.

In such a violent, purposeless world, it is hardly surprising that for some people even an illusion of security is better than nothing. Where else did Ajax the tortoise have to go? Security is not a definable, quantifiable state: it is a perception of the many different factors relating to one's self-interest. People have not yet realized that the concept of security encompasses far more than simple military force, let alone that additional military expenditure is actually undermining their security. The Brandt Report puts it like this:

Few threats to the peace and survival of the human community are greater than those posed by the degradation of the biosphere on which human life depends... In the global context, true security cannot be achieved by a mounting build-up of weapons (defence in a narrow sense), but only by providing the basic conditions for solving the non-military problems which threaten them.[5]

The threats we face today are less likely to arise from a breakdown between nations than from the breakdown between humanity and the Earth. The national security of countries all over the planet is already threatened by the destruction of the biosphere, yet it still seems impossible for them to realize that these threats outweigh traditional military threats. By spending money on the latter, they exacerbate the former; by maximizing their military strength, they are in fact undermining their national security.

In a *Guardian* article, Dr Norman Myers explained this point:

Today's 'threat situations' are not amenable to traditional responses in the form of military initiatives. In essence, they are far from situations of the 'What I gain, you lose' type. Rather, they are situations of the 'We all gain together, or we all lose together' type. Hence, there is no scope for established modes of response, lying with the competitive assertiveness of individual nation-states. Rather, the responses must lie with co-operative endeavour on the part of nations acting together.[6]

For instance, a country which did not base its economics on the indiscriminate pursuit of growth would have considerably eased the pressure on world resources. And it is these diminishing resources that will prove the flashpoints of the future. In the same article, Dr Myers warns us:

As more and more people seek to sustain themselves from fewer and fewer resources, we can anticipate a period of inevitable shortages, disputes, and armed conflicts.

Lasting peace can be based only on a genuine understanding of the relationships between people and planet. At the moment, the very opposite is the case. To most world leaders, peace means *pax economica*, something to be achieved exclusively through

economic and industrial development. Such development invariably means the transformation of subsistence-oriented societies
into reluctant clients of the world economic system. This in turn
creates conditions of scarcity, empowers new elites, undermines
self-reliance and destroys local customs. It promotes violence
against the environment and positively encourages the destruction of the global commons. There is nothing peaceful about *pax
economica*!

Ecologists insist therefore that one cannot talk about peace in a
vacuum; it must be related to one's way of life, to one's spiritual
concerns and to the rest of one's policies. Frankly, given the
policies of the other parties in the UK, particularly as regards
economics, trade, resources, nuclear power, the environment and
the Third World, for them to claim that they are promoting
peace is an outrageous lie: they are actively and consistently
promoting policies that make peace absolutely impossible. Of
necessity, industrialism begets belligerency. The green movement's prime responsibility must be to make other peace
campaigners aware of the connections between the politics of
industrialism and the likelihood of war. Our message to the rest of
the movement is clear: peace is indivisible.

Keeping the Peace

A vital part of 'making the connections' is to understand the need
for significant reductions in military expenditure. The cost of
military spending today precludes any possibility of a peaceful
transition to a sustainable society. It absorbs resources, skills, cash
and scientific expertise at such a rate as to deny us all a future.
Although the problems of the global arms trade must eventually
be solved at the international level, we should demonstrate our
commitment both to human rights now and to freedom in the
future by withdrawing from the arms trade as soon as possible.

This will of course require a massive conversion programme.
There's been a lot of general discussion about such a programme,
but as yet very few concrete plans. Many of the alternatives are
certainly very exciting, and many point in the direction of a
sustainable society: more efficient transport development, solar,

wind and wave technology, heat pumps, fuel cells, new construction techniques, health equipment, machine tools and recycling
devices. There's no question that these are things people need (far
more than they need new weapons or refinements to Concorde!),
but needs do not necessarily translate into orders in our 'free
market' industrial system. It's not really a technological problem
we face, nor indeed a logistical one (in the eighteen months after
the end of World War II, 8 million workers were redeployed
from military to non-military production), but a political problem.

Which brings us back full circle to the question of relative
threats, and how we comprehend the meaning of security. It
remains a question of the hearts and minds of ordinary people,
and it has to be said that the Cold War propaganda of this
Government has seized many of those hearts and minds in a grip
of steel. We now find ourselves in the ludicrous situation where
many people have come to believe that peace is threatened by the
strength of the peace movement! And that security can only be
guaranteed by paranoia, hatred and fear! In such a context, peace
means nothing more than the absence of war, and the weapons of
mass destruction on which we depend may fittingly be called the
'peacekeepers'.

One of the hardest things to face up to is that the roots of war
often lie deep within ourselves. From the games we played as kids
(and now buy for our children) to the television we watch as
adults, a web of violence permeates our lives. This is what Johann
Galtung refers to as 'structural violence', which extends even to
the paying of taxes for military purposes, to the unuttered
violence of our thoughts, and the spoken violence of our
communication. Peace begins in these secret places, in demilitariz-
ing ourselves, in rejecting violence and, while acknowledging the
justification for armed resistance, in denouncing the glorification
of war. War never was and never will be glorious.

Peace can arise only from developing relationships of trust and
mutual acceptance. The fear and the paranoia must be removed
before we will have a chance to make meaningful reductions in
today's arsenals. We must reject the cynicism and apathy that
makes it so much easier for people to dig their own graves rather
than cope with the implications of not digging them. As Jim

Garrison says, 'We must learn to extend to the Russians the simple recognition that our common humanity unites us far more powerfully than our differences divide us.'[7] Without lasting peace, visions of a sustainable and harmonious balance between humanity and our planet are mere pipe dreams. But peace itself is a forlorn hope if we cannot simultaneously achieve that balance. Many people throughout the world are realizing that the only answer to this crisis is not to try to persuade the ruling powers to give up their authority and change their methods, but to build a new world from the grass roots, geared to peace, freedom, justice and the satisfaction of human needs.

12
Sustainable Society

Human Scale

Since 'small is beautiful' is about the one green slogan to have penetrated the mind of our mass industrial society, you might suppose that advocates of green politics would live, breathe and swear by it. Unfortunately it has become one of those catch-all panaceas that are obsessively extolled by some greens, greatly to the detriment of clear thinking and useful action.

Human scale (the title of an absolutely gigantic book by the American author Kirkpatrick Sale) is a far more useful starting point. For there is no one single appropriate size for organizations, institutions or groups of people: different structures are appropriate for different purposes, and what matters is the sense of *balance*. Fritz Schumacher himself was well aware of the need not to over-react:

> For constructive work, the principal task is always the restoration of balance. Today, we suffer from an almost universal idolatry of giantism. It is therefore necessary to insist on the virtues of smallness – where this applies. If there were a prevailing idolatry of smallness, irrespective of subject or purpose, one would have to try and exercise influence in the opposite direction.[1]

Appropriate scale depends on three factors: the extent to which people can *identify* with the structure of the group concerned, and thereby overcome some of the alienating tendencies in modern society; the extent to which they feel in *control* of their own involvement, so that the question 'What can I do about it?' becomes largely redundant; and the degree to which they have a chance to exercise that control *responsibly*. It means matching institutions, technology and social structures to our human scale, rather than being obliged to adapt to an inhuman scale.

What is big, therefore, is not necessarily bad: it's when *all* things are big that it's bad. And what is small is not necessarily beautiful: history abounds in examples of cruel, barbarous, uncivilized, narrow-minded smallness. In our own time, there can be just as much oppression and incompetence in a small unit as a big one, and a small country can run industrially amok just as thoroughly as a big one. As usual, Roszak puts his finger on it: 'The opposite of big is not small, but personal... the problem of scale is finally not in our institutions, but in ourselves.' If any organization or any group is vitiated by the blighted ethic and practice of industrialism, it makes not a ha'p'orth of difference whether it's big or small.

Appropriate scale means we must think both big and small at the same time. Greens rightly place a lot of emphasis on living in smaller, more independent communities, which will require shifting a lot of today's economic and political activity downward to a more local scale. The stress here is on the principle of *diversity*. But we should be aware that the growing interdependence of all people on this planet means that the next stage of social evolution will be a move towards some sort of cohesive planetary civilization – and here the emphasis is on the principle of *integration*. It should be clear by now that from a green point of view, these two principles are mutually dependent; we are unlikely to achieve the one without the other.

People Power

This dual emphasis on decentralization and internationalism is quite unique to the green perspective. It defies accusations of insularity and little Englandism: there can be no such thing as unilateral ecology. In our pursuit of greater security, lasting prosperity and a better quality of life, it seems all too obvious to us that only in smaller, more self-reliant communities will people feel that sense of loyalty and involvement which allows for the full development of individual potential – and with modern communications technology, there need be no fear of a return to the mean-minded parochialism of pre-industrial Britain. And it seems equally obvious that a sense of loyalty to the planet is a precondition for our survival. 'Act locally, think globally' remains the most useful slogan for what we're trying to achieve.

In terms of restoring power to the community, nothing should be done at a higher level that can be done at a lower. The question is, how do we decide whether or not it can be done at a certain level, and who makes the decision? For all its merits, our democracy ensures that ordinary people do not have the *power* to exercise personal responsibility; they are encouraged to place the ultimate responsibility for things outside themselves, and in blaming the 'system' or waiting for things to be done on their behalf, that is exactly what most people seem content with. Talk of greater self-reliance and doing more for ourselves goes down like a ton of bricks on many an inner-urban doorstep! The *representative* element of the system has insidiously undermined the element of *participation*, in that turning out to vote now and then seems to have become the be-all and end-all of our democracy. Green parties throughout Europe do not reject representative democracy, but are determined that it should be complemented by a more personal and participatory form of politics – and that means looking at power in a different way. James Robertson puts it like this:

> The new power will be seen as the absence of dependence, and as the ability to help others to shake off dependence... instead of seeking to overpower the adversary, the non-violent approach seeks to withdraw power from the adversary. Less powerful people are less powerful because they have been conditioned to give power to more powerful people and organizations.[2]

Such a concept of power is not dependent on political parties or on Parliament, for it seeks change through transformation rather than confrontation.

Again, it is important to work out some sort of balance here. To suppose that such a transformation will just come about of its own accord is extraordinarily naive even by green standards. I shudder to think of the number of initiatives that have wasted away, the amount of idealistic energy that has been squandered, and the whole gamut of opportunities that have been lost because of foolish regulations, a lack of financial support, and every conceivable form of political and institutional obstruction. The taking of power from below, by this process of self-empowerment, *must* be combined with the passing down of power from above. The voice of transformation must also be heard and be influential within the existing system – otherwise we're just wasting our time. Without conventional political activity to bring about the appropriate legislation, the balance of power will never be significantly altered.

Hence the need for devolved assemblies in Scotland and Wales, and a move towards regional assemblies throughout England. Parliament would still retain many of its powers (concerning defence, trade, finance, resource management and pollution, for instance), for that is the appropriate level to deal with such matters; but the 'regionalizing' of the UK, and the far-reaching devolution of certain powers to county or district level, is an essential element of a sustainable society. As I mentioned in chapter 7, such reforms would have to be accompanied by the introduction of proportional representation at every level of government. PR is *not* a panacea for all our problems, and there will be nothing necessarily ecological about a Parliament or a council just because it happens to have been elected by PR: if all they continue to represent are the vested interests of industrialism, it doesn't much matter what proportions they come in.

However, it's a start. And there are many other areas in which legislation does indeed have a crucial role to play, such as women's rights and racial equality. Despite considerable progress, we still find certain groups discriminated against and exploited. Equal rights, equal pay, equal job opportunities, equal treatment in law: these are ideals that still need to be converted into realities.

But we've all had enough experience of social legislation of this sort to know that it can never do the job it's intended to do without corresponding changes in values and attitudes. Decentralization depends as much on people accepting their personal and community responsibilities as on specific measures of devolution. Racial equality depends as much on blacks and whites working out a shared sense of direction and co-operation within the community as on existing legislation to prevent discrimination. 'Law and order' depends more on the elimination of poverty and inequality, on closer links between community and police, than on new laws or tougher penalities. And good health depends more on people leading interesting, balanced lives, with fulfilling work and a healthy diet, than on high-technology, high-cost medicine.

Healthy Mind, Healthy Body

The provision of health care is a classic example of the way in which social responsibility should be divided between the Government and the individual. It's hardly surprising that the National Health Service is on the point of collapse, since it allows so little scope for the individual to take responsibility for his or her own health. Furthermore, as Fritjof Capra points out, we seem to have forgotten 'the interdependence of our individual health and that of the social and ecological systems in which we are embedded'.[3] In a sustainable society, the Government would finance a health service with the emphasis on preventive medicine, smaller, community-based hospitals, and a far greater degree of regional autonomy. Money currently spent on relatively high administrative costs and the extravagances of high-tech medicine would be reallocated to health education and neighbourhood health-care schemes. In each locality, some doctors' surgeries would be expanded into Environmental Health Agencies to operate these schemes, and alternative health-care methods would be encouraged and promoted.

That would be the Government's side of the deal; ours would

be to follow its advice as regards good health and to keep *ourselves* healthy. Given that it's more important to save people's lives rather than to raise revenue by encouraging ill-health, such a Government would probably ban all cigarette adverts and all smoking in public places – but it would still be up to individuals to decide for themselves what risks to take with their health. Given that it's quite clear that many cancers are related to diet, such a Government would ban many more potential carcinogens and provide detailed information about the dangers of processed food – but it would still be up to individuals to decide for themselves how much junk food they wished to poison themselves with. A Government cannot force adjustments to a collective lifestyle; it can show where the path to good health lies, but it can't make us jog down it.

There is already a powerful grass-roots movement involved in promoting healthier lifestyles and emphasizing the importance of personal responsibility. The recent establishment of the British Holistic Medical Association is indicative of the trend away from 'reductionist' medicine and towards the idea of seeing each person as an integrated whole, a totality of mind, body and spirit. This inevitably involves profound changes in our value system and social organization; holistic medicine, the bringing of wholeness to alienated individuals and fragmented societies, is perhaps the most profound of the forces for 'cultural renewal' that are already beginning to flourish in industrial societies.

It is sadly not possible to detect the same potential for renewal within the world of education. Our schools are still entirely fashioned by the demands of an urban industrial society, promoting competition, suppressing individuality, pushing the old line of getting good grades to get good qualifications to get a good job – but suddenly the jobs aren't there. What may once have been fine as an ideal and more or less fair in practice is now neither. And there's no new ideal to take its place. The trendy spasm of fervour for the notion of 'deschooling' has rightly sputtered out; we're all aware of the 'injustices of compulsory miseducation', but that should not serve as a pretext for pseudo-radical rhetoric about abolishing schools. However threadbare it may sometimes appear, the pattern of social justice and equal opportunity in the

UK is rooted in the provision of compulsory education. There'd be plenty of privately run schools and 'skills exchanges' in Kensington, Hampstead and Islington, but not so many in Hackney, Newham and Brixton.

So we're stuck with schools – and there's much progressive, dedicated work going on in those schools up and down the country, for all the high-pitched criticism of the anti-comprehensive, right-wing rump. Radical reform *is* necessary, though as a teacher myself I know that the prospect of change and yet more change can be disheartening: teachers and parents alike are often dwarfed by the contrast between what happens and what we'd like to happen. What can we do, for instance, about promoting the ideal of education for life, so that people of all ages have lifelong access to a continuing process of learning and self-discovery, and are enabled at any time to enlarge the options open to them in a rapidly changing world? What can we do to make our schools more community-oriented, with fewer children and smaller classes (for small is certainly beautiful in that particular context!), to turn our secondary schools into community colleges, offering courses to people of all ages and providing many essential community facilities? What can we do about changing the curriculum so that instead of focusing on a redundant exam system, it combines the teaching of essential skills with the promotion of humane, pluralist values, and provides access to the ways of the world and current affairs while allowing each child's individual interests to flourish?

Some parents choose to withdraw their children so as to organize their schooling themselves, and this is a precious right that must be carefully protected. But most of us just lump it; parents don't ask too much, and teachers try to avoid becoming cynical or dream about setting up their own schools! And some, knowing that the values and materialistic attitudes of industrialism are on the turn, bide their time and create a microcosm of a better world in their own classroom. In this respect, teaching *is* a subversive activity. But teachers and parents alike must find some way of organizing themselves if holistic education, *education for life on Earth*, is to have any significant bearing on the future.

Village Life

Without such changes in health and education, our cities will remain fairly desperate places. And yet here we should not be biding our time; the 'greening of the city' has already started, admittedly involving only tiny minorities as yet, but they are sowing acorns as they go. Even planners are beginning to realize the importance of flexibility, to understand that rigid notions of the 'ideal built environment' matter far less than the web of social and human relationships that give life to that environment. The single greatest reform would be to change the planning laws, so that communities could again be built up around a cluster of diverse activities, with workshops, light industry, housing, the neighbourhood school, health-care facilities, community gardens and the local shops and pub gathered together in a great higgledy-piggledy planners' nightmare. Through a much stronger sense of identity and loyalty one could establish genuine 'urban villages', each one becoming more self-reliant in terms of local production catering for local needs. With more people working from home or within the community, we could at the same time save ourselves from ghastly office blocks and the scourge of 'development', while liberating people from the strain and expense of commuting.

Ivan Illich has drawn an important distinction between 'enabling' and 'disabling' professions. We have of late suffered from a surfeit of the latter, but here too things are changing. Officially or unofficially, architects and planners have been involved in all sorts of spontaneous initiatives in the rehabilitiation of old streets, co-ownership schemes, and the setting up of workshops or urban farms. Instead of relying on the old token participation, they are working with the people involved right from the start. There is certainly plenty to be done as regards the maintenance of existing housing stock, restoring run-down properties to create higher-standard accommodation, providing work for small builders and self-employed skilled workers, and promoting housing associations, co-operatives and self-help groups.

Just another ecotopian dream? Possibly – but I'd be interested to know what else is going to happen to our cities. If anything, ecology matters more in an urban context than it does in the country, though some people still think that all ecologists are born wearing a pair of wellington boots! As the most unsustainable feature of an unsustainable system, cities pose something of a challenge to those intent upon sustainability – and yet, despite the problems, there are enormous opportunities to create a more convivial and caring way of life.

Such a process will be greatly assisted by the continuing drain of people from our cities. There are probably millions who would like to leave, but can't because they're tied to jobs they don't like or houses they can't afford. These people must be helped to extricate themselves from their predicament as part of a massive programme of rural resettlement. People look a bit embarrassed when such a notion crops up, as if one were talking about the compulsory establishment of a new peasant culture! There need be nothing compulsory about restoring the balance between rural and urban. Over the last thirty-five years Governments have paid out billions in subsidies and grants to promote a way of farming that has caused millions to move away from the countryside. As our rural communities have withered away, our cities have become bloated beyond any reasonable notion of human scale. A bit of thinning down is in order, and there's no reason why a different use of grants and subsidies should not reverse the process; simply by raising the number of people working on the land to the average level for all EEC countries, we would create more than 1 million jobs! To achieve this we would need radical measures of land reform, but as an interim solution many people have suggested that local councils could buy up suitable large farms that came on the market and re-sell them in smaller units, thus restoring much needed vitality to the countryside.

There's much more to rural regeneration than getting people back on the land, and we need to shift our attention to what has been called 'sporadic development', bringing back old crafts and encouraging new skills. Existing housing stock should be renovated in preference to new building, but there's no reason why marginal land should not be used for this purpose as long as it is in harmony with the environment. You can't regenerate the countryside without people, and you won't get the people without providing somewhere for them to live and work. There's a lot of justifiable concern about the possible encroachment on farming land that such a resettlement programme would entail: but little prime farming land need be lost, and much marginal land or derelict land would be better used as a result. It would certainly be an improvement on the depopulated and dehumanized wasteland of industrial agriculture.

From the Ashes

The face of Britain would be radically changed as a result of these changes: a better balance between town and country would mean healthier lifestyles, more self-reliant communities, huge reductions in transport costs, and a rapid return to the principles of good husbandry. This in turn would have an enormous impact on energy demand and would allow us to develop a very different kind of energy strategy.

'The great dream many of us had thirty years ago that nuclear energy would set us free has been turning to ashes': such are the words of Alvin Weinberg, one of the founders of the nuclear energy programme in America. The dream is indeed over. Like a recurring nightmare, the unpalatable truths about nuclear power have cropped up throughout this book: far from being clean, safe and cheap, nuclear power produces small amounts of the wrong kind of energy at considerable expense and with considerable risks. From start to finish, the nuclear cycle creates more costs than it does energy. At the start of the process, the mining and milling of uranium has caused terrible damage to people's health and to the environment. We would never tolerate in Britain what the multinationals have done in the name of progress in places like Namibia. And at the end of it, all the reactors are still there, silently leaking radioactivity; they either have to be dismantled at enormous cost or left there for centuries, deadly monuments to this generation's greed and irrationality. Nor should one suppose that one reactor is any better than another. The fast-breeder reactor will be even more expensive, will entail vast quantities of plutonium moving around the country, and is totally irrelevant in the context of substituting for oil and gas. The attempt to develop fusion energy is the last lingering convulsion of a technocratic nightmare.

There's only one answer, and that's to start decommissioning the monsters *now*, and to close down all research establishments apart from those dealing with nuclear waste. The immediate closure of all the nuclear power stations would still leave us with a considerable surplus of generating capacity. It's therefore fundamentally dishonest for politicians to go on and on about the dangers of nuclear waste and the risks to health and yet remain committed to the continuation of nuclear power. Don't they realize that by allowing the CEGB to hang on to the tattered shreds of its plan for an all-electric economy, the UK has no real energy policy at all? In a country where the opportunity for change is almost unique, where technological opportunities are greater than ever before, and where the imminent depletion of North Sea oil stares us all in the face, we are still stuck in the nuclear mire. Colin Sweet:

> nonsense cannot give way to common sense so long as the nuclear fantasy continues to addle the minds of those in power. The result is a confusion. Britain does need an energy policy, but it can only get it by an act of decision which removes the unreal option of nuclear power and allows others to be considered.[4]

At the very most, we have ten years to put this country's energy requirements on a sustainable footing before supplies of oil and gas fail to meet demand. It's a perfectly feasible challenge, on two conditions: first, that energy demand continues to decline; and secondly, that any such strategy is based on the distinction between 'capital' and 'interest' in terms of energy sources: the so-called 'energy crisis' will be caused not by any shortage of supplies, but by a grotesque and self-defeating appetite for continual expansion. A low-energy strategy means a low-consumption economy; we *can* do more with less, but we'd be better off doing less with less. Energy prices must therefore rise to reflect true costs, our capital resources must be strictly conserved, and our energy needs must be increasingly met by the efficient use of solar, wind, wave, tidal and bio-mass energy, which is inexhaustible, abundant, non-polluting and safe.

Energy for the Future

The most important energy source of all is conservation –energy saved is always cheaper than energy generated. In our wasteful, energy-rich society there is incredible scope for reducing demand; Gerald Leach and his colleagues at the International Institute for Environment and Development calculated that the widespread promotion of existing technologies would keep energy demand constant even if levels of GNP trebled, thus achieving a threefold increase in energy efficiency.[5] New technologies will make a massive contribution to this process; some techniques (such as insulation and heat recovery systems) are already more than cost-effective; others will become so as soon as relative energy costs rise. As well as thermal insulation (which could reduce space heating demand by 40 per cent), we should expect to see considerable savings through more sophisticated controls of heating systems, improved efficiencies in boiler design and performance, the use of heat exchangers and controlled ventilation systems.

It really all depends on the Government using its influence and vast range of powers to promote conservation. In this respect, our current Government is pathetic; it's not even particularly good at pious exhortation, which is the usual stock in trade of Governments. We need hefty grants and incentives for insulating new buildings and retrofitting old ones; thermal efficiency standards should be introduced, so that the need for low energy consumption is incorporated into our building regulations; and energy tariffs need to be turned on their head. At the moment, the more you use, the cheaper you get it; what we should do is sell it cheaply up to a certain level (to safeguard low-income groups) and then at progressively higher rates after that to encourage conservation.

In the UK 16 per cent of energy consumption is used on travel, and there's plenty of scope here for reducing consumption. Only the greenest of the green actually talk about doing away with motor cars, but none of us is keen on encouraging them. No more motorways would be built and there'd be restrictions on motor cars in most cities. In the future we're going to be terribly handicapped unless we can shift most of our freight on to the railways, investing heavily in electrification and the improvement of industrial rail links, and shift most of our transport needs from private to public transport. There needs to be much stronger incentives to choose vehicles with better m.p.g. rates, and there's no reason why the road tax should not be set at different rates accordingly. If people have got to drive, they might as well do so as efficiently as possible! None the less, it won't be long before we have to start saving our oil for other chemical and industrial purposes. We should already be slowing the rate of extraction from the North Sea, gradually raising the price, and using every single pound we get out of it to ensure a smooth transition to a more sustainable society.

In the meantime, our reserves of coal should see us through for the next 250 years or so. This will be the basis of our energy supplies during the transition period. There *are* risks and high environmental costs associated with this, but modern power stations using fluidized bed combustion methods could double their efficiency standards and eliminate most noxious emissions. The biggest challenge confronting the coal industry will be the development of combined heat and power stations, whose waste heat will be used for district heating schemes. This is already widely practised on the Continent, but the CEGB (surprise, surprise!) has never been interested, since it considers that its exclusive role is to generate electricity, not to use energy wisely.

It is, of course, one of the great advantages of renewable sources of energy that it's much harder for the monopolies and multi-nationals of this world to manipulate their supply as a means of making money. This will be very much to the advantage of all Third World countries, as well as to domestic consumers in this country. Once you've installed your solar panels, you have a constant, independent, all but free source of energy. The potential for solar power in Britain is still enormous – and our climate really isn't the handicap that people make out. Between May and September the quality of sunlight is quite adequate for water heating and space heating. According to the UK section of the International Solar Energy Society, as much as 12 per cent of the UK's primary energy could be provided by solar power before 2020. As usual we're a long way behind other countries, and much work still has to be done on appropriate heat-storage systems, but the growing interest in solar architecture and passive solar design is most encouraging. As energy analyst Christopher Flavin observes: 'Even assuming substantial growth in housing, the world's buildings may be using 25 per cent less fuel and electricity in the year 2000 than they do today – an important step toward achieving a sustainable world energy economy.'[6]

Millions and millions of words have been written about the potential of renewable energy sources, and this is not the place to try to recapitulate them all. Suffice it to say that with the right levels of research and investment, Britain could be the world leader in renewables. Clusters of windmills deployed off-shore in the North Sea, a tidal barrage across the Bristol Channel, wave-power generation off the north coast of Scotland, small water turbines up and down the land – this is where our money should be going. Of course there are problems, and of course it's expensive – nor will all such development necessarily be in the 'small is beautiful' mould. But we're talking about the difference between a reasonable future and or a very squalid one.

Close to the hearts of decentralist ecologists are developments in the use of marginal land for fuel crops and energy plantations. Photosynthesis is nature's way of achieving sustainable growth, turning solar energy into fixed energy, though one should never forget that using land in this way means it can't be used for growing food. We should also be making far better use of the waste that our society generates so prolifically, either through direct combustion to provide district heating schemes or through conversion into alcohol or methane. Methane digesters, producing both energy and organic fertilizer, provide a significant proportion of China's energy, putting to the best possible use all those agricultural wastes, livestock manure and even human wastes. Sustainability means making use of everything: it's no good turning up our noses at so convenient an energy resource.

And in the very process of becoming sustainable, we would be creating hundreds of thousands of jobs. In his *Low Energy Strategy for the United Kingdom*, Gerald Leach wrote:

> The emphasis on conservation would create a great diversity of jobs, unskilled as well as skilled, in thousands of factories and workshops across the country – in sharp contrast to the specialized

and limited job opportunities implied by conventional supply-expansion energy forecasts and policies.[7]

We would at the same time be producing goods with a considerable export potential in every corner of the world. And that's the final irony about it all: one of the reasons why the CEGB is pushing ahead with the PWR is because of its lingering hopes of developing an export market for such reactors. I don't suppose there's a single soul outside the CEGB who seriously expects that sales will ever exceed zero. Yet again, we're turning our backs on the future.

13
A Green and Pleasant Land

The Wisdom of the Land

Farmers today are very clever; with the help of the Government and the Common Market, they continue to make a great deal of money from what they do. But they are not wise, for they are serving their own interests only by working against the interests of the land. In the long history of farming, the post-war period stands alone in its attempt to create prosperity despite the Earth, rather than through the Earth. The balance is already beginning to swing back, and with the passing of the Age of Oil, most of the worst abuses will necessarily be halted; we shall reaffirm the principles of good husbandry by working with the land to provide good work, good food and stable rural communities.

Most Governments are not known for having their ear to the ground and this one is no exception. I wonder if there are any in the Ministry of Agriculture to whom the Earth whispers its warnings, suggesting that it might be better to use the vast array of grants and subsidies to encourage conservation rather than to bribe farmers to make a lunar landscape of our countryside; that smaller farms are more economic than large ones, though the latter may *appear* more profitable; that mixed, rotational farming is more productive and more sustainable in terms of soil fertility than monocultural production; or even that the use of organic methods can be as profitable as conventional ones?

Food production should be encouraged at every level of society, not just on farms, but on smallholdings, city farms, in allotments, back gardens – even window-boxes. Small-scale, labour-intensive methods are by far the most efficient way of growing food. We've recently seen a sudden surge of interest in what is called 'permaculture', short for 'permanent agriculture', pioneered by an Australian called Bill Mollison. Permaculture has four basic requirements, as explained in an article in the *Ecologist* by Penny Strange: it must produce more energy than it consumes; it must not destroy its own base through misuse of soil or water resources; it must meet local needs, not serve some mass-produced, processed and packaged market; and it must find all the necessary nutrients on site, without depending on inorganic fertilizers. Its success depends on very careful design, the use of a very large number of plant and animal species, the recycling of all materials, and hard work. It allows for the intensive use of small areas of land, and seems to me to be especially important as regards food production in cities.

In short, permaculture is a 'self-sustaining, cultivated eco-system'; its implications for a change in the attitude of people to the planet are highly significant. It binds people to the natural processes of the Earth and, with the use of appropriate technology, creates a sense of harmony that is sorely lacking. Such an approach is doubly attractive in that it encourages self-reliance and personal responsibility, as well as being the wisest possible use of the environment, both rural and urban. Because our urban environment is so horrendously polluted, the idea of cities growing their own food is currently quite impossible. What with the lead and all the other toxic emissions of cars and industries, the average London-grown lettuce is a positive health hazard. It seems abundantly clear that the 'polluter pays' principle should operate in this as in many other cases; if car owners create the pollution, they should pay for it by being obliged to fit catalytic converters or other pollution controls.

The record of this and previous British Governments over measures of environmental protection is an international disgrace; time after time the only reason why anything happens is because the Government is obliged to take action because of EEC directives. It's unlikely that there will ever be much improvement in this lamentable state of affairs until a decision is taken to set up an Environmental Protection Agency – not least to protect the environment from the Ministry of the Environment! It goes without saying that such an Agency would have to have real teeth, with considerable powers of inspection, regulation and punitive fining. But the environment has never been a political issue in this country; nor, for a long time, has the whole question of land reform. It's almost as if the politicians had decided among themselves never to discuss the issue: the Conservatives are doing very nicely (most of the Cabinet are landowners themselves); Labour can only bleat on about nationalizing the land, which everyone knows is a non-starter; and the Liberals have given up trying to get across the ideas of Henry George. And that's a pity, for it strikes me that the only way to break the monopoly of landownership would be the introduction of some form of land tax. Though the individual ownership of houses, farms, workshops or anything that improves the usefulness of the land would continue, it should not be possible to own the land itself. People should be tenants of the land rather than owners; instead of our present rating system, a Community Ground Rent, assessed according to the value of the land, should be paid annually to the community. Nobody would want to pay the Community Rent on land they were not using, so it would cease to be an investment proposition, and land speculation could be brought to an end. The benefits of the land would derive solely from sustainable, ecological use. If we accept that the land is part of our common wealth, then part of the wisdom of the land is to ensure that many more people have access to it than they do today.

Natural Wealth

Richard St Barbe Baker, the greatest of all tree lovers, once said: 'A nation's wealth, its real wealth, can be gauged by its tree cover.' That leaves us fairly and squarely on the poverty line, for whereas Germany has 30 per cent tree cover and Italy 27 per cent, we have a mere 9 per cent tree cover. And when you think that we import about 90 per cent of our timber products, at the cost of more than £3 billion a year, that's a pretty extraordinary deficiency! Every single long-term ecological and economic forecast predicts a shortage of timber by the turn of the century.

If we're to do anything about that, we must consider both the demand and the supply of forestry products, and basically follow the same rule as with energy: reduce demand and increase the quality of supply. A reduction in demand will certainly result from the different pattern of economic development that is becoming apparent. Once we've thrown out our throw-away values, and opted for inconspicuous consumption, we shall be using far less timber and paper products anyway – with the one exception that far more wood will be used as a source of fuel in wood-burning stoves. The supply side is more complicated. Even with a considerable extension of present reafforestation programmes, we still wouldn't be any more than 25 per cent self-sufficient by the year 2025 at present rates of consumption. Many people, including environmentalists, oppose such programmes anyway, given that reafforestation today seems to consist exclusively in laying down drab carpets of conifers all over the country. But with different patterns of agriculture, allowing for more shelter belts, extended coppicing and small-scale plantations, and an approach to large-scale reafforestation based on both hardwoods and many different kinds of softwood, we could work towards the possibility of getting this country's tree cover up to about 30 per cent *without* causing further damage to the environment.

One excellent proposal appeared in the January 1980 issue of the *Ecologist*, suggesting that a Forestry Bank should be set up; using money deposited with the Bank by pension funds and other City institutions, farmers would be encouraged to plant more trees by receiving an annual return rather than having to wait for a lump sum at the end of the forest's natural cycle. The trouble about such a scheme is that it demands vision, clear thinking and a concern for the future; such attributes are rarely to be discovered in the nooks and crannies of a political system that can't think beyond the lifetime of whichever Government happens to be in power. Hence there has been very little support for it. In much the same way, it is patiently explained to ecologists that most paper-recycling schemes are just not 'economic'. In other words, there's no profit in them.

Ironically, the more sophisticated a society is today, the more dependent it becomes on others for goods it no longer makes itself. An affluent white minority in an overwhelmingly non-white world, much of it poor beyond our imagination, can hardly expect to have the continuing benefit of other countries' cheap raw materials. The UK's record on the recycling of raw materials and consumer waste is therefore particularly dismal: 10 per cent of steel, 3 per cent of glass containers, 20 per cent of paper products. Nearly 90 per cent of all consumer waste is just buried in landfills. We are unbelievably profligate, and though all politicians hate to propose a new tax, the only solution would seem to be a resources tax, operated as a tariff on all imported materials and as a direct levy on indigenous production. Such a measure would reduce imports, reduce overall consumption, encourge more efficient re-use, recycling and repair, promote new investment and research, and guarantee more responsible treatment of industrial and domestic waste. Together with a Minimum Packaging and Container Deposit Act, such measures would both reduce the amount of waste and ensure that it wasn't actually wasted.

One has to be blind to the whole trend of the world economy not to realize that inflation, increased demand and shrinking supplies are going to bring about such changes *whether we like it or not*. Again, it's a question of preparing now for the inevitability of tomorrow. To render obsolete the very notion of built-in obsolescence, to use less energy and fewer resources in the very process of creating more jobs: these are goals that even a short-sighted Government might quite easily embrace. But even as we begin to move in this direction, don't suppose that such reforms will be sufficient in themselves. The fiction of combining present levels of consumption with 'limitless recycling' is more characteristic of the technocratic vision than of an ecological one. Recycling itself uses resources, expends energy, creates thermal pollution; on the bottom line, it's just an industrial activity like all the others. Recycling is both useful and necessary – but it is an illusion to imagine that it provides any basic answers.

Whatever Befalls the Earth

'Basic answers' are sometimes rather hard to live with. Imagine
the surprise of the US Government in 1855 when it received this
response from Chief Seattle of the Dwamish Indians to a proposal
to buy some of their land:

> How can you buy or sell the sky? We do not own the freshness of
> the air or the sparkle on the water. How then can you buy them
> from us? Every part of the Earth is sacred to my people, holy in
> their memory and experience. We know that the white man does
> not understand our ways. He is a stranger who comes in the night,
> and takes from the land whatever he needs. The Earth is not his
> friend, but his enemy, and when he's conquered it, he moves on.
> He kidnaps the Earth from his children. His appetite will devour
> the Earth and leave behind a desert. If all the beasts were gone, we
> would die from a great loneliness of the spirit, for whatever
> happens to the beasts happens also to us. All things are connected.
> Whatever befalls the Earth, befalls the children of the Earth.

'Whatever happens to the beasts happens also to us.' That's a
genuinely radical premise to work on, but it explains why many
green activists are so deeply involved in upholding the basic rights
of other species. For us, it is not enough to protect animals for
practical, self-interested reasons alone; there is also a profoundly
moral concern, rooted in our philosophy of respect for all that
dwells on this planet. In the short term that means that the live
export of farm animals for slaughter should be banned, voluntary
codes on animal rights should be made mandatory, all imports
into the UK of furs and skins and products deriving from
endangered species should be prohibited, no experiment should
be carried out on animals without an anaesthetic, and the use of
animals for *all* tests on cosmetics, for tobacco and alcohol
research, and in weapons or biological and chemical warfare
programmes should be outlawed immediately. In the longer
term, vivisection would be abolished, all hunting and coursing
with hounds would be banned, battery farming would be phased
out, our reliance on animals to meet our need for food would be
reduced – and *then* we could start living in harmony with the rest
of creation!

It sounds like a tall order, but few would actually suffer in the implementing of it, and we would all benefit intangibly. The strident voices of those who make their living from the suffering of animals should be heeded, for there is in such a clamour a chilling reminder of the sickness at the heart of our industrial culture. But they should not be obeyed. No more should the clamour of those farmers who seek to profit from the unnecessary destruction of the flora and fauna of our countryside, for the same reminder is to be found there. Sites of Special Scientific Interest and National Parks must be given full legal protection, and the relevant sections of the Wildlife and Countryside Act repealed. As Friends of the Earth have suggested, this country urgently needs a comprehensive Natural Heritage Bill to protect our countryside properly.

Both in this country and internationally, the stress needs to be on the conservation of whole habitats rather than species-by-species campaigns. The logic of the latter approach has always been suspect: you save one species, and move on to the next; some you win, some you lose, but nothing changes concerning the basic pressures that threaten them all. At the end of the line, there'll only be ourselves to save. That was why the success of the Tasmania Wilderness campaign was particularly gratifying. It was a classic confrontation: in one corner, the Tasmanian Government, already with a surplus of electricity on its hands, but keen to build a new dam to attract new industry; and in the other, those who wanted to save the unique flora and fauna of one of the world's three remaining temperate forests. In 1982 it became only the thirtieth site to receive UNESCO's 'World Heritage' status, but this meant little to the Tasmanian Government, whose premier described it as a 'leech-ridden ditch'! Mutual incomprehension was the inevitable result, resolved only by the enlightened intervention of the Federal Australian Government.

Just imagine, if the Tasmanian Wilderness had gone under, what sort of example that would have set Third World countries, which face infinitely greater problems than the affluent Tasmanians in finding a balance between development and conservation. And I wonder how the Government of Tasmania would respond to this proposal from Erik Eckholm to ease the burden of Third World countries in this respect:

Because so much responsibility for preserving the Earth's genetic
heritage falls to poorer countries, the possibility must be considered of distributing the costs of conservation fairly among
nations. If the world's extant species and gene pools are the
priceless heritage of all humanity, then people everywhere need to
share the burdens of conservation according to their ability to do
so.[1]

Its answer would, I imagine, be brief, and along the lines that
the issue had nothing to do with it. Such a response would be
typical of the Government of every single developed nation. At
UN conference after UN conference we have seen Governments
happily voting in favour of all sorts of international reforms; but
it's rarely that anything ever gets done to put such resolutions into
practice. The sense of genuine interdependence which resulted
from the first wave of ecological concern in the early 1970s has
since become sadly atrophied. In retrospect, the 1972 UN Conference on the Human Environment was the high-water mark in
terms of planetary awareness and solidarity.

One World...

Only One Earth, the title of Barbara Ward's influential study on
global politics, remains only an ideal. In the face of prolonged
recession, concern for the environment and concern for the Third
World are now considered rather 'wet', having little to do with
the hard realities of politics. The global commons carry no weight
against the onslaught of sovereign interests; stirring up nationalist
or chauvinist sentiments will do much more for your political
career than preaching moderation on behalf of the planet.
Nationalism has thus become an integral part of the industrial
world order, as nation competes with nation for slices of a
dwindling economic pie. And yet we should remember that
nationalism was not always dominant in world affairs, and that
the need for renewed internationalism has never been greater.
Without it there is no conceivable chance of establishing any new
international economic order. 'The principle of national
sovereignty is one of the major obstacles to the collective

salvation of humanity':[2] such was the conclusion of Aurelio Peccei.

We must therefore encourage the development of the world into a confederation of sustainable communities at every turn. We must strengthen the peacekeeping and collective security role of the UN, and while pressing for the reform of their often extravagant and bureaucratic procedures, and the broadening of a rather limited vision, we should continue to support the other agencies of the UN, and ensure that they too are moving in the right direction. The responsibility for taking such initiatives undoubtedly lies with us, for it is we who are the arch-exploiters of the Earth's resources, and the arch-manipulators of the world's poor.

Consider, in such a light, the uselessness of the Treaty of Rome. By all appearances, such a treaty should be promoting the cause of internationalism, and yet it serves to do exactly the opposite. Each member state is in there using it for its own advantage; the interests of the *whole* of Europe, let alone the rest of the planet, hardly get a look in. Moreover, the Treaty of Rome has to be one of the most unecological documents ever written: the EEC is totally committed to industrial growth and expansion, to nuclear power, to the economies of scale, to mass consumption, and to a Common Agricultural Policy that is wasteful, expensive and destroys the soil. While the bureaucrats and the plutocrats thrive in every European capital, the European Parliament impotently tries to do something about its own impotence. Over-centralization ensures that social and regional policies are quite inadequate, and the Lomé Convention perpetuates the exploitation of parts of the Third World in the name of international solidarity.

It's hardly surprising that greens want the UK to quit the Common Market. But we still wish to remain closely linked to Europe, seeing the way forward through a federation of regions rather than a squabble of nation states. The Treaty of Rome should be scrapped, and a Treaty of Larzac signed in its place. Larzac is in a tough and inhospitable region of south-west France, which recently saw a memorable triumph of the local people over the French military machine. The army wanted to extend its already sizeable military training area, but met with strong opposition from the peasant farmers who didn't want to lose their land and livelihood. The farmers refused to pay their war taxes, collected the money themselves and used it for non-violent productive purposes. They 'grazed' their sheep under the Eiffel Tower to gain sympathy and publicity, and in the end the army had to back down. A treaty signed at such a place would emphasize the principles and practice of non-violence, direct democracy, the interests of decentralization and regionalism, the rejection of militarism, the need for self-reliance and sustainability, and the right kind of agriculture! Such will be the attributes of the new internationalism, with one major addition: massive effort to help the people of the Third World.

...For the Third World

For ecologists, such a commitment doesn't just mean making available a higher proportion of our GNP; it means alternatives which totally reject the traditional model of development. Those who might regard this country's withdrawal from international markets as 'undesirable isolationism', and an evasion of our responsibilities to the Third World, should reflect that international trade has merely seen the rich grow richer and the poor poorer, lending support to intolerant regimes, and destroying local economies and native cultures through the imposition of an utterly inappropriate Western industrial way of life. When there's growth in the North, the South is pushed into an unsustainable pattern of development, losing out at both ends of world trade. When there's no growth in the North, the South is still tied to us through the politics of the begging bowl. In an international system based on greed, mistrust and domination, one person's growth is another's famine.

The rhetoric of most Third World leaders concerning a new international economic order is quite redundant. We are witnessing the extraordinary process of 'self-colonization', for their obsession with exports and world trade is an obvious hangover from colonial times. Essentially, they must find ways of gradually de-linking their economies from those of the developed world, and building up more sustainable patterns of trade between themselves. This does not mean we can just forget our obligations to those countries that are trapped on the world trade roller-coaster, and there would have to be exceptions to any import control policy which we ourselves pursued. An interesting way of meeting our obligations, while simultaneously encouraging self-reliance, would be to refund them with the proceeds of any tariffs raised on imports from their countries for the specific purpose of diversifying their economies to escape the import/export trap.

There is, of course, a paradox here. To attain the kind of interdependence I have been talking about Third World countries must first achieve *independence*! Just as there can be no healthy integration without the widest possible diversity, so it will be impossible to establish new principles of interdependence until each and every country is able to stand on its own two feet in terms of meeting the basic needs of all its people. Such an understanding of self-reliance and sustainability is not likely to be particularly attractive to many Third World leaders who are doing very nicely out of the present system. Aid (in the form of grants, not loans) should be given as an incentive for promoting the right kind of growth. Wholesale land reform to control land speculation, to provide security of tenure and to ensure that the disintegration of rural life is stemmed will be a vital part of this process. The emphasis must be on growing food for domestic use, not for export, with the aim of becoming entirely self-sufficient in food production. And we must promote a pattern of development that uses appropriate technology to provide meaningful employment for people without destroying the environment. There is no clash between ecology and this sort of development; indeed, the two are dependent on each other. Sensitive development is essential if we are to preserve the biosphere, and attention to soil conservation, watershed management and forestry is essential if we are to safeguard people's real quality of life.

A classic example concerns the way in which we can help Third World countries to meet their forestry needs. The World Bank is already shifting the pattern of aid away from large-scale industrial timber to what is known as 'community' or 'social forestry'. If increased tree-planting for firewood is to be successful, radical changes in the role of national forest agencies are called for. Participation and community involvement are not just pious aims, but practical necessities; it's not just a question of managing the forest to ensure a financial return, but rather of providing the skills and management techniques to ensure that the local people are the main beneficiaries. High technology has little part in such a concept of progress: the most important reform of all would be to distribute cheap, efficient wood stoves, which would have the effect of reducing firewood needs by more than 50 per cent.

None of which alters the uncomfortable fact that with demand exceeding sustainable yield, if some are going to consume more, others are going to have to consume less. The basic, unmet needs of the poor can be met only by a reduction in the consumption of the developed world. The basic problem for the Third World remains the First World – that's us. Should we not therefore write off many of the debts that are crippling Third World countries, and making the transition to a more sustainable future quite impossible? Should we not impose rigorous controls on the operations of multinational companies, particularly concerning the repatriation of profits, transfer pricing, dumping, patent rights and working conditions? And above all, should we not recognize *for ourselves* that helping the Third World is not just a question of giving more, but of taking less? Which brings us right back to the voters of Croydon and the need to promote a simpler, less materialistic way of life. We could *all* aim to eat more simply, travel more cheaply, live less wastefully and cut out the horrors of conspicuous consumption.

Population Concern

In terms of reducing overall consumption, there's nothing more effective than reducing the number of people doing the consuming. The slight drop in the birth rate of the UK is therefore encouraging; it reflects not only long-term social changes (such as improvements in eduction and in employment prospects for women), but also the choice of ordinary people deciding for themselves how to achieve the optimum quality of life. This instinctive choice at the individual level needs the strongest possible reinforcement at the national level, especially since many of our laws and social institutions still exert a powerful pro-natalist influence.

For some time now, Population Concern and other interested parties have been calling for a proper national policy on population. People need to see that the Government of this country is thinking of the future and realizes the implications of our remaining so overcrowded. Inspired leadership on this issue is of the greatest importance, far more so than any specific legislative changes. Some people argue the advantages of getting rid of family allowances and child benefits, but these are now so low that it's impossible to suppose people are swayed one way or the other by their existence, and there's certainly no case to support the fantastical notion that people are having more children specifically to squeeze money out of the Chancellor of the Exchequer! Having babies is an expensive business, and it's going to get more expensive. None the less, there may be a case, as much for symbolic and psychological reasons, in doing away with such payments after the second child, especially if that made it easier to find additional sums of money to promote better education for parenthood in all secondary schools and improvements in the Family Planning Service. There are still 200,000 unwanted pregnancies every year, of which about half end in abortions. Good preventive health care means we cannot afford penny-pinching when it comes to the provision of comprehensive, enlightened family planning.

Population size is also influenced by the number of people coming in and out of the country. The strictly logical position, as far as ecologists are concerned, is to keep immigration at the lowest possible level while remaining sensitive to the needs of refugees, split families, political exiles, etc. In an already overcrowded island, such an approach seems both legitimate and rational. It is also part of the logic of ecology that such an approach should *in no way* be discriminatory in terms of race or colour. One is hardly likely to promote the kind of internationalism I have been talking about without an uncompromising commitment to the equality of all races.

A widely debated, humane and realistic population policy would have a considerable impact on people in this country. It would also make it far easier for us to participate actively in helping Third World countries to reduce *their* population. We've seen that even with fertility rates falling to replacement levels, the world population will continue to grow for several decades: hence the World Bank and UN projections of between 10 billion and 12 billion people by 2025. To me these figures seem horrifying and ridiculous – a measure of our inability to realize just how serious the ecological crisis has already become. We should be aiming to stabilize world population at a level well below that, at around 6 billion or 7 billion at the most. This means that fertility may well have to drop below replacement levels, with one-child families becoming the norm.

Such a goal will require a huge effort, and much of the financial wherewithal must come from us. For too long now there has been an inconclusive debate between those (mostly on the left) who believe that economic and social development is the key to reducing population, and those (mostly on the right) who emphasize family planning. The development v. contraception debate is in fact quite futile, since the one depends on the other: a lack of development hampers the effectiveness of family planning schemes, and a lack of family planning makes it impossible to achieve any development anyway! The sad fact is that for many countries to wait for development would be suicidal. Family planning services must therefore be made available to all who need them as soon as possible, and should be backed by legal abortion. One-third of the world's women are still denied this basic right. At the same time, better health care and education remain absolutely crucial components of any population policy; people must be enabled to understand that what may be rational in one way (having more children because of high infant mortality or to provide security in old age or additional labour) is desperately irrational in many other ways.

Few other countries have fully accepted this challenge. The experience of China, whose population exceeded 1 billion in 1980, is an interesting exception. Realizing the implications of its declining population/resource ratio (it has to subsist on $\frac{1}{4}$ acre per person), it has enthusiastically promoted the one-child family, providing significant increases in monthly salary for those who take the pledge, as well as many other benefits and preferential measures. The success China has had has been possible only because of widespread public discussion and awareness, though there is now considerable concern that in a few of the more remote areas of China, cultural conditioning still puts so high a value on male children that baby girls are being killed at birth.

It remains vitally important to reject coercive measures as an unacceptable and morally repugnant infringement of human rights, but many countries are boxing themselves in by hoping that the problem will just go away. In this instance, the costs of procrastination rather than action are appalling to contemplate, and for millions will mean the difference between survival or extinction. Unless we take action now, this sort of dilemma may well become commonplace in the overpopulated world of the next century.

14
Spirits of the Future

Determining Human Nature

Today's dominant social paradigm offers little scope for compromise and systematically represses the articulation of alternatives. Yet with the basic needs of the majority of people on this planet still unmet, with widespread poverty still a part of almost all developed nations, with alienation an endemic feature of the way we organize ourselves, it is clear that the system has failed even in its own terms. Worse still, the 'Faustian bargain' we have made has resulted in a totally distorted set of values and a complete loss of any spiritual dimension. The cohesion of industrial society is ensured by an all-embracing alliance of vested interests, including manifestations of so-called 'opposition', subsumed within the super-ideology of industrialism. Economic values dominate the deliberations and decisions of our democracy; human, non-material values are readily sacrificed in the pursuit of material power. The future is seen to offer no more than a continuation of the present, a deterministic acceptance of a world system that is clearly breaking down.

This determinism is rooted in a particular and ominous view of human nature. When all rational argument fails, critics of the green approach desperately fall back on the old line that we're just naive idealists, that what we propose is simply not possible when confronted with the 'reality of human nature'. Human nature is held to be immutably aggressive, intolerant, self-seeking and shallow, and will therefore work against any initiatives to move towards a sustainable future. Such critics refer us back to our own analysis of the destructive, scorched-earth history of humankind, and triumphantly conclude by claiming, 'You can't change human nature.'

This little catch-phrase, the self-fulfilling prophecy of all determinists, often tells us more about their particular brand of cynicism, apathy, helplessness or alienation than about human nature! There is not a shred of evidence that we are genetically modelled to behave for ever as we do today, but the belief that it is possible to change depends upon a belief in free will and the potential in each of us to counter the deterministic programming of contemporary society. As Duane Elgin says, 'Like the seed with the potential of becoming a flower, human nature is not a static thing but a spectrum of potentials.'[1] In today's wasteland much of the human potentiality for good is inevitably thwarted, but we should never ignore the creativity, the capacity for vision and compassion and the vast resources of moral energy that are innate within each of us. There is nothing naive about realistic idealism: without it, the pattern of the future will be a sad and sorry affair. No one has expressed this more clearly than R. H. Tawney:

> It is obvious that no change of system or machinery can avert those causes of social malaise which consist of egotism, greed or quarrelsomeness of human nature. What it can do is to create an environment in which those are not the qualities which are encouraged. It cannot secure that men live up to their principles. What it can do is to establish their social order upon principles to which, if they please, they can live up and not live down. It cannot control their actions. It can offer them an end on which to fix their minds. And, as their minds are, so in the long run and with exceptions, their practical activity will be.[2]

Self-interest is and will remain a fundamental characteristic of human nature; but in today's world individual interests are identified almost exclusively with the accumulation of material wealth, and few politicians are prepared to articulate a concept of social progress that is not totally dependent on increases in GNP. The problem for the future is to ensure that the interests of the individual are more in line with those of society at large and with those of the planet. The materialistic ethic of mass consumption has even managed to obscure the ultimate goal of survival, and has obscured it so successfully that survival and the pursuit of individual self-interest, in the manner prescribed by *all* politicians apart from the greens, are now mutually exclusive.

It is not so much the political process itself that will determine our fate as the values on which that process is premised. 'Values are the key to the evolution of a sustainable society not only because they influence behaviour but also because they determine a society's priorities and thus its ability to survive.'[3] That is the uncompromising message throughout Lester Brown's work, and he goes on to quote these words from US environmentalist Tom Bender: 'Values are really a complex and compact repository of survival wisdom – an expression of those feelings, attitudes, actions and relationships that we have found to be most essential to our well-being.'[4]

Wants and Needs

Society's values are neither timeless nor absolute; they change according to our changing circumstances and our perception of these circumstances. The dominant values of industrialism are already under the microscope, and many will necessarily be rejected as we move towards a more sustainable society. There will, for example, be far more attention paid to the distinction between *wants* and *needs*, needs being those things that are essential to our survival and to civilized, humane existence, wants being the extras that serve to gratify our desires. We all need good food; some people want to subsist on a diet of extravagant and often harmful luxuries. We all need to get from A to B; some people insist they can manage such a feat only in the back of a Rolls-Royce. We all need clothes; some people aspire after a new outfit for every day of the year. The distinctions are not always cut and dried, and the manipulative skills of the advertising industry in converting wants into needs make it difficult to expose the excesses thrown up by mass-consumption industrialism. By today's standards, keeping up with the Joneses is a positive social virtue.

To distinguish between genuine needs and artificial needs requires an unequivocal value judgement, the distinction between the two depending on the extent to which their satisfaction *genuinely* contributes to our well-being. Is our well-being genuinely enhanced by electric toothbrushes or umpteen varieties of cat food? Indeed, is it genuinely enhanced by having 'unlimited' freedom of choice? There are no easy answers; we are merely insisting that such questions must now be put, for continued reliance on the operations of a so-called 'value-free' market amounts to no more than a cowardly and perverse refusal to face reality. That reality tells all of us, whatever our political allegiance, that we must renounce the suicidal imperative of *more* for the sustainable logic of *better*.

The work of the psychologist Abraham Maslow suggests the direction in which we should be moving. He put forward the idea of a hierarchy of needs, suggesting that once our basic survival or physiological needs are met, 'higher-order needs' become increasingly important. Once we have food, clothing and shelter, we concern ourselves more with satisfying human relationships and the ways in which we experience a sense of belonging. The next 'level' (I find it difficult to interpret the hierarchy too literally) involves the need for recognition, for status, social position and self-esteem; and finally we move on to what Maslow refers to as the need for 'self-actualization', making the most of the multifarious talents and creative resources with which we are endowed. The problem, of course, is that in our alienated, materialistic society these higher-order needs are often 'satisfied' in an alienated, materialistic fashion: people may seek to '*buy* love', to '*win* a circle of friends', to '*acquire* status', to '*gain* self-respect'. But there is no binding imperative that tells us that these needs *have* to be met materialistically; indeed, it is part of *our* understanding of human nature that they may be met both more rewardingly and more sustainably in non-materialistic ways. If we are to redirect the drives that lie behind these needs towards less materialistic and destructive goals, then many of our current symbols of success and achievement must be altered. Henryk Skolimowski quotes Denis Healey's plaintive words: 'What people want are stable

prices and a secure job. These things aren't very exciting to
visionaries, but they are what most people want, and it is very
difficult to get them. Trying to get them is not an ignoble thing to
do.'5 Indeed not, but nor is it adequate, whether or not you see
yourself as a 'visionary'. Consider, by way of contrast, the four
'consumption criteria' proposed by the Simple Living Collective
of San Francisco:

1. Does what I own or buy promote activity, self-reliance
   and involvement, or does it induce passivity and
   dependence?
2. Are my consumption patterns basically satisfying, or do I
   buy much that serves no real need?
3. How tied is my present job and lifestyle to instalment
   payments, maintenance and repair costs, and the expectations of others?
4. Do I consider the impact of my consumption patterns on
   other people and on the Earth?6

I'd lay odds that many of you are now thinking, 'How naive,
quite unrealistic, hardly the stuff of *real* politics.' Just remember
that those, like Denis Healey's, are *value* judgements, the product
of an obsolescent value system that couldn't look reality in the
face even if it knew where to find it. And just remember the
words of J. M. Keynes: 'The power of vested interests is vastly
exaggerated compared with the gradual encroachment of ideas.'

Metaphysical Reconstruction

Only now are we beginning to realize the urgency of transcend-
ing our industrial perspective, of discovering new values and new
ways of relating to each other and to the planet. Schumacher
referred to this as a process of 'metaphysical reconstruction', and
greens today see this in terms of at least four components: the
person, the people, the planet and the spirit. The rest of the
chapter looks at each of these in turn, reconfirming the obvious
but often neglected truth that politicians are powerless unless they
move with the spirit of the times. We believe that spirit has as

little in common with the weary reiteration of collectivist abstractions as it does with the mean-minded promotion of individualist self-concern.

We prefer to talk of a 'third way', of the *politics of the person*, in the belief that only a completely different approach can resolve today's paradox of scale: that things are both too big and too small at one and the same time – too big because we are all made to feel like pygmies, too small because we are incapable of adopting a genuine planetary ethic. By stressing the importance of personal responsibility, by refusing to accept that any of us are neutral in our actions or decisions, green politics enables people to determine appropriate responses in a complex and confusing world. Barry Commoner once wrote: 'Like the ecosphere itself, the peoples of the world are linked through their separate but interconnected needs to a common fate. The world will survive the environmental crisis as a whole, or not at all.'[7] The acceptance of human interdependence must be at the heart of any new ethic: in Hazel Henderson's words, 'Morality has at last become pragmatic.'

The development of that sort of planetary consciousness depends upon our being able to rediscover our links with the Earth, and to work in sympathy with rather than against the organic harmonies that make life possible. This is the most important feature of what ecologists refer to as 'holism', embracing the *totality* of something in the knowledge that it is so much greater than the sum of its component parts; things cannot be understood by the isolated examination of their parts. The wisdom of the future depends on our ability to synthesize, to bring together rather than to take apart. So often the total picture, the sense of the whole, either eludes us or is wilfully set aside. Politics has reduced the 'average voter' to an opinion-poll readout of immediate material needs; science has reduced the planet to a quantifiable aggregate of physical resources. It is the job of ecologists to re-present the whole picture, in all its diversity, complexity and frustrating unquantifiability!

To do so we rely on an extraordinarily eclectic political and philosophical ancestry. To try to weld this into some easily articulated ideology really would be a waste of time – and would completely miss the point. Ideologies are by definition both reductionist *and* divisive. And ours, sadly, is a society that seems to thrive on divisions: East and West, North and South, mind and body, them and us, black and white, winners and losers. By choosing to live the way we do, we have ensured that the wholeness of each individual is crushed beneath the divisive pressure of industrialism, that the interdependence of all living creatures is constantly disregarded in the name of 'economic necessity', and that the oneness of humanity is deliberately denied through a suicidal obsession with national sovereignty. It is the wisdom of ecology that now begins to re-instruct us about the crucial importance of balance and the holistic interrelatedness of life on Earth. Metaphysical reconstruction means the making whole of each one of us.

Reclaiming the Feminine

The artificial splits that exist today are a reflection of a culture divided against itself and separated from the natural world by delusions of its own superiority. Any sense of wholeness is all but impossible unless we are first able to rid ourselves of this dominant world view. *All* radical movements must understand this imperative. The greens have always stated that the challenge which faces people in the modern industrial state is something much deeper than the struggle for political or economic rights. The problem goes right back to the mechanistic world view of industrialism, and the way in which its dominant values repress all of us, men and women alike.

Hand in hand with the exploitation of the Earth has gone the continuing social, economic and political repression of women in particular. For all the many advances and liberal reforms, this remains an undeniable feature of contemporary society. Women are still oppressed and exploited, and daily exposed to injustice, violence and discrimination. In times of high unemployment women are always affected more than men, and even when they do have 'equal opportunities' in terms of employment, there is never a doubt that they are the ones who are expected to carry out most of the responsibilities of parenthood and looking after the home.

This is the most noticeable aspect of what is rightly referred to as our 'patriarchy', defined by Adrienne Rich as a 'system of power in which men determine what part women shall or shall not play, and in which the female is everywhere subsumed under the male'. There is much that can and must be done to bring an end to such visible injustices. But there are other aspects of our patriarchal system that are harder to deal with, for the repression of women is carried through by men who for the most part are themselves the repressed victims of a patriarchial world view. I am talking now of the different qualities and energies which coexist in *all* of us, but which have become so dependent on sexual stereotyping as to leave the vast majority of people repressing a whole side of their nature.

There is a real problem about defining these qualities. Some people refer to them as 'masculine' and 'feminine', insisting that they should not be linked with notions of 'male' and 'female'. Others use the terms 'yin' and 'yang', but not only is this somewhat esoteric, it's also the case that such terms are usually dangerously distorted. You find a certain type of writer implying that all yin (or 'feminine') qualities are good, and all yang (or 'masculine') qualities are bad. In Chinese culture yin and yang have *never* been associated with moral values. As Fritjof Capra says: 'What is good is not yin or yang, but the dynamic balance between the two; what is bad or harmful is imbalance.'[8] We can achieve wholeness only by achieving balance between what I think of as the tough, harder qualities of human nature (such as competitiveness, assertiveness, the rational and analytical, the materialistic and intellectual) and the gentler, or softer qualities (such as co-operation, empathy, holistic thinking, emotion and intuition). Balance means exactly what it says, combining the two in mutual harmony. It is therefore not appropriate for people to claim that green means exclusively soft/yin/feminine, for to do so demonstrates a fundamental failure to comprehend the import-ance of balance. *No one* has the right to usurp the green way to suit his or her own particular imbalance!

That said, we must accept that we live in a world in which all semblance of such balance has been lost. Patriarchy means nothing less than compulsory masculinity. Positive moral values *are* conferred on masculine qualities, while feminine qualities *are*

consistently invalidated, denigrated and suppressed. As far as men
are concerned, there's a closed shop when it comes to emphasizing
qualities 'traditionally associated' with the opposite sex, and this is
ruthlessly reinforced both in schools and in the home. In so
distorted a world women have been forced into an impossible
choice: either they identify with their stereotyping, in which case
they are expected to accept passive, submissive roles, or they
reject it and compete with all the rest of the toughies, in which
case they are criticized as aggressive and unfeminine! Both
stereotypes are the product of patriarchal industrialism, both are
unbalanced and correspondingly impoverishing. Women who
have freed themselves from these stereotypes have provided
inspiring models for the liberation of men, suggesting ways in
which all of us must work to become more authentically what *we
already are*.

What is clear is that women are more subtly and intensely
aware of the 'dis-ease' of contemporary industrialism caused by
this lack of balance. They can see that the crisis is fundamentally a
spiritual one, as the violence done to people, to other species and
to the planet spreads inexorably. While remaining aware that this
has often been used as a prison in the past, it is important today to
re-emphasize the age-old identification of women and nature,
that consciousness of how all living forms are interrelated in the
cyclical rhythms of life and death. I am powerfully reminded of
these links by the campaign still going on in New Zealand to
allow Maori women to claim their afterbirth from the affronted,
super-hygenic administrators of New Zealand hospitals. The
Maoris have the same word (*te whenua*) for land and for afterbirth,
and the link is there not only in the language but also in the
traditional custom of burying the afterbirth, making it one again
with the Earth. Maoris believe that the Earth is a kind of
'elemental womb', from which we are all delivered and to which
we all eventually return; this consciousness of the Earth as part of
our spiritual heritage has been sadly eroded by the immersion of
the Maori people in a predominantly white, materialistic and
patriarchal culture. It is highly significant that in many social
and political campaigns it is now the Maori women who are
showing their usually dominant menfolk that there's nothing
particularly dignified or fulfilling in meekly adopting the status

of second-class citizens in a fundamentally misguided society.

These women have literally 'joined forces with the life of the planet and become her peculiar voice'. So too, I believe, have the women of Greenham Common. For all the grotesque distortions of the media, it has been their example that has inspired many of us to find that extra ounce of commitment, and to discover within ourselves where the path of peace lies. Those extraordinarily ramshackle shelters outside the perimeter fence symbolize far more than a bastion of opposition to Cruise missiles; the songs they sing are far more than a way of keeping up their spirits. They are a challenge to the whole patriarchal order of society, and their commitment to 'putting their bodies where their beliefs are' has done much to illumine the legal and moral underpinning of that society. Over and above the fact that this Government's policy is almost certainly illegal in terms of the Nuremberg principles, the Geneva Protocols of 1977 and the 1969 Genocide Act, the Greenham women's courageous, non-violent direct action has reminded us that there are times when the dictates of our conscience *must* transcend the laws of the land.

There has, of course, been much controversy about whether they are right to insist that actions at Greenham should be on a 'women-only' basis. In the short term this inevitably causes some frustration to the rest of the peace movement, but its long-term value may well be of far greater significance. For it is true (let me at least acknowledge it in my own case, for all that I endeavour to do something about it!) that many men in the peace movement carry with them much of the baggage of their patriarchal orientation. None the less, there are limits to exclusivity: the potential is there within *all* of us to discover that peace depends on restoring the balance both within ourselves and between ourselves and the Earth. This is not the prerogative of women – it is the duty of us all. The women's movement will undoubtedly be a major force in any cultural transformation, perhaps even providing the links that will allow other movements and interests to flow together, but its message must be articulated in a way that embraces all humanity, women and men alike. I would suggest that this remains impossible without a green understanding of the indivisibility of all life on Earth.

When Less is More

In terms of our actual lifestyle, the holistic principle of *balance* means achieving a sufficiency, finding the middle way between indulgence and poverty, doing more with less. Many refer to this as 'voluntary simplicity', a phrase that up until now has carried more weight in the USA than in Europe. Duane Elgin's excellent book of the same title defines voluntary simplicity as a way of life that is 'outwardly more simple, inwardly more rich', with the emphasis more on human relationships and personal development than on material consumption.

For many years now there has been evidence of substantial support for the principle of treading more lightly on our fragile planet. The famous Harris survey of 1977 demonstrated, in its own words, that 'Significant majorities place a higher priority on improving human and social relationships and the quality of American life than on simply raising the standard of living.' Some of the findings really were quite startling: for instance, 76 per cent of people questioned expressed a preference for 'learning to get our pleasures out of non-material experiences' rather than 'satisfying our needs for more goods and services'. Since then this 'quiet revolution' has been confirmed in many studies both in the USA and in Europe; on the basis of the work done at the Stanford Research Institute, Duane Elgin reckons that 20 per cent of Americans may now reasonably be described as 'inner-directed consumers', 70 per cent as 'outer-directed consumers' (i.e. those motivated by the goals of conventional materialism) and 10 per cent as 'need-driven consumers', made up of those on or near the poverty line.

In the mid-1970s a major poll in Norway indicated that 74 per cent of people would prefer a 'simple life with no more than essentials' to a 'high income and many material benefits if these have to be obtained through increased stress'. This poll was linked to a movement called The Future in Our Hands, which has been campaigning since then for a new lifestyle and a more equitable distribution of the Earth's resources. No such poll has been carried out in the UK, though in many ways the work of those involved in the Life Style Movement parallels the Scandinavian organization. There is a clarity of message and intent about the Life Style Movement that makes it extremely appealing, and it is sad that it has not grown more rapidly. 'To live more simply that all of us may simply live', 'to give more freely that all may be free to give', 'to recapture our lost humanity we must recall our sense of oneness with the whole Human Family' – these are the simple truths of the future, and comprise the only ethic that allows us to reconcile the genuine needs of each person with the needs of all humankind and of the planet.

Voluntary simplicity is ecology in action: it's difficult to be an ecologist without working towards some such set of principles. It is not just the latest trendy manifestation of ex-hippies and hippies *manqués*, for it is a response to a series of crises that are *already* transforming our lives. The push of necessity thus combines with the pull of opportunity to find a better way of living. Though some may talk of 'joyful asceticism', the emphasis is *not* on sacrifice but rather on the enormous benefits to be derived from such a way of life. Nor should it in any way be confused with 'dropping out' or turning one's back on progress: voluntary simplicity will provide the very bedrock of progress in the future, with its emphasis on mutual support, non-violence, human scale, self-reliance and self-determination.

The roots of such a movement reach deeply into the needs and ideals of millions of people. The search for materialistic fulfilment is essentially self-defeating, for it is a search that is permanently directed away from the self. It's horrifying to reflect on just how much time we spend in the *involuntary* pursuit of superficial gratification; voluntarily to pursue a process of inner growth and learning may sound pretentious to some, but only because the balance between inner and outer has been so grievously distorted. It is tragic that in all our Western arrogance we seem to have learned so little from the meditative traditions of the East, whose concept of spirituality and personal growth is so radically different from our own. For a start, there is room within them for awareness of the planet: it is the cornerstone of Taoism, for instance, that only by living in harmony with the natural environment is it possible for people to develop their own creative potential. The Buddhist concept of *esho funi* emphasizes a relationship of mutual dependence, in which people and their environment are seen not as two separate entities but as one indivisible unity. Buddhists would contend that the predominant Western assumption that nature exists for the exclusive use of humankind can be overcome only in the context of a new spiritual awakening.

Gaia

From this point of view, quite the worst part of the *World Conservation Strategy* is the section on 'Environmental Ethics and Conservation Action'. It is premised on the statement that Man (its terminology, not mine!) 'is both apart from Nature and a part of Nature'. The latter is self-evident, but on what grounds can one defend the former assertion: our use of speech, the invention of religion, our superior intelligence, or perhaps our temporary dominance? The belief that we are 'apart from' the rest of creation is an intrinsic feature of the dominant world order, a man-centred or anthropocentric philosophy. Ecologists argue that this ultimately destructive belief must be rooted out and replaced with a life-centred or biocentric philosophy. The *World Conservation Strategy* has done its best to sit on the fence on this crucial issue, with painful consequences for both the authors and the whole environment movement: an environmental ethic so lacking in integrity can serve only to reinforce the destructive power of industrialism today. For our survival depends on our being able to transcend our anthropocentrism.

No more powerful contribution has been made to this debate than James Lovelock's fascinating book *Gaia: A New Look at Life on Earth*. His observations of the ways in which the Earth's basic life-support systems are maintained have led him to believe that the Earth 'constitutes a single system, made and managed to their own convenience by living organisms'. How is it, he asks, that the amount of oxygen in the atmosphere is maintained at just the right level, when with either too much or too little life would not be possible? How is it that the balance of oxygen and methane 'inexplicably' remains constant? How is it that the temperature at the surface of the Earth has been maintained at a comfortable level for billions of years (give or take the odd ice age!) despite the fact that the heat reaching us from the sun has increased by as much as

40 per cent? These things are far too convenient to be mere coincidences; the planet, it seems, is controlled by its living organisms, and these organisms behave as if they were a single entity, actively shaping the conditions for life on Earth. 'We share the chemistry of all the non-humans among which we live; everything that lives on Earth is made of the same stuff.' Lovelock refers to this as the 'Gaia hypothesis', after the Greek goddess of the Earth.

Were such a hypothesis to be 'proved', it would certainly put the kibosh on any lingering anthropocentric fantasies! But it would also affect the outlook of many conservationists, for Lovelock stresses that Gaia should not be interpreted as some munificent mother figure protecting us from harm; there is no 'purpose' behind such a system, and it is all so vast that the extinction of any one species is neither here nor there. Yet one can't help but be struck by the way in which his hypothesis is so clearly linked to an older and once universal philosophy that experienced the Earth as a divine being, as *Mother Earth*. 'The Earth is a goddess and she teaches justice to those who can learn, for the better she is served, the more good things she gives in return.' Xenophon's words express the belief of the ancient Greeks that they were the children of the Earth, and that the Earth was an animate, living organism with a 'natural law' which rewarded those who protected her and struck down those who harmed her. (Remember Orion, who claimed in his arrogance that he could kill all animals on Earth, only to be rubbed out by a monstrous scorpion!)

The ancient Greeks and Romans saw the Earth in much the same way as the Red Indians. It is worth quoting this extraordinary extract from John Neihardt's recordings of Black Elk, first published in 1932. Black Elk is explaining the importance of the sacred pipe:

> Before we smoke, you must see how it is made and what it means. These four ribbons hanging here on the stem are the four quarters of the universe. The black one is for the west where the thunder beings live to send us rain. The white one for the north whence comes the great white cleansing wind; the red one for the east, whence springs the light, and where the morning star lives to give

men wisdom; the yellow for the south, whence comes the summer, and the power to grow. But these four spirits are only one spirit, and the eagle feather here is for that one, which is like a father, and it is also for the thoughts of men that they should rise high as the eagles do. Is not the sky the father and the Earth a mother, and are not all living things with feet or wings or roots their children? And this hide upon the mouthpiece here, which should be bison hide, is for the Earth, from whence we come and at whose breast we suck as babies all our lives, along with all the animals, and birds and trees and grasses. And because it means all this, and more than any man can understand, the pipe is holy. Hear me, four quarters of the world – a relative I am. Give me the strength to walk the soft Earth, a relative to all that is.

There was nothing sentimental about this sense of 'kin relationship', but it perfectly expresses a tradition that has always run parallel to the mechanistic conception of nature that we looked at in chapter 8. In many writers and thinkers, nature becomes a metaphor for God; the Earth itself becomes the sacred centre of life. In Wendell Berry's moving book *The Unsettling of America*, we see the extent to which our separation from nature is the source of contemporary alienation, and the way in which harmony with nature is central to any new spirituality. The Earth speaks to something in all of us, and each of us needs some special place where we can hear the song of Gaia, where the 'living continuity of person and planet' becomes again an effortless part of us. It's so much easier for me to write that now, surrounded by trees that I have planted, on a small piece of land that possesses me rather than the other way round, than it would be in London – and such is the agony of many an ecologist today.

I hope that doesn't sound too mystical, for we need to reassert the unity of humankind and nature without necessarily relying on quasi-religious concepts. Reverence or respect for the Earth should suffice. The Greenpeace philosophy states: 'Ecology teaches us that humankind is not the centre of life on the planet. Ecology has taught us that the whole Earth is part of our "body" and that we must learn to respect it as we respect ourselves. As we feel for ourselves, we must feel for all forms of life – the whales, the seals, the forests, the seas.' It is in that spirit (if I may so loosely use the word) that thousands have responded to the stirring example of the Chipko Andolan, the 'hug-the-trees' movement, founded by Chandi Prasan Bhatt in the foothills of the Himalayas in 1973. Originating with the women of one small village who became concerned at the devastating effects of deforestation, the Chipko movement has now spread its campaign to protect the forests throughout northern India and to other small Himalayan states; backed by fasts and demonstrations, inspired by many of Gandhi's non-violent principles, the villagers have refused to become the victims of so-called 'development'. Their movement has profoundly influenced government forestry policy, leading to the growth of community forestry and many decentralized, forest-based cottage industries. Extensive reafforestation of the right kind of trees will ensure self-sufficiency in forest needs and sustainable employment for many who would otherwise leave the hills to seek employment in the cities. 'Ecology is permanent economy': that's their slogan, and one that seems appropriate for all good Gaians!

Only Connect!

'And God said, Let us make man in our image, after our likeness: and let them have dominion over the fish of the sea, and over the fowl of the air, and over the cattle, and over all the Earth, and over every creeping thing that creepeth upon the Earth.' What a lot of problems that little passage from Genesis has caused! For 'dominion' has usually been interpreted as 'domination', which has in turn provided a licence for the wholesale exploitation of the Earth. Christianity has often participated in suppressing a sense of reverence for the Earth, despite the 'pagan' hangover of the Harvest Festival.

One of the most crucial tasks for Christians today is therefore to reinterpret the meaning of 'dominion' in terms of stewardship and ecological responsibility for life on Earth. But this is only the beginning, for it must be acknowledged that the established Churches have got themselves into a fearful pickle. Though it must be obvious to all Church leaders that today's spiritual vacuum derives from the ascendancy of scientific materialism, they have so tamely accepted that religion and politics are two separate things that they are all but incapable of making any useful contribution to resolving today's appalling problems. I speak here of the Church as an *institution*, and mean no offence to the many *individuals* within those Churches who are in the forefront of social and political change. To put it bluntly, is it any wonder that most people ignore the Church, when on the one hand it professes to follow the example of Christ, and on the other refuses to take a stand against weapons of mass destruction, against the redundant 'Protestant' work ethic, or against the assaults of materialism? As Erik Dammann says in *The Future in Our Hands*:

> How can we go about our daily business as if the rest of the world did not exist? How can we sit in our churches and talk of love? How dare we teach our children about justice and humanity without doing anything at all for those 82,000 children who are starving to death *every day*? How can we, without cringing, without even thinking of cringing, set an increase of consumption as a goal for ourselves, when we know what this means for others?[9]

Tough questions, I admit. But I can't help but be astonished at the sheer lack of urgency among Church leaders today; ours is a world crying out for leadership, for some kind of spiritual guidance. And yet, as the winds of change whistle up their richly caparisoned copes, where on earth are they? Can't they see that thousands are already engaged in rediscovering the essence of Christian spirituality, without excessive concern for orthodoxy or dogma? Can't they see the green shoots creeping up between the flagstones of their deserted cloisters? (I ask these questions and make these criticisms as a hopelessly imperfect Christian, who has great difficulty with many parts of the Creed, but one of whose greatest joys is growling his way through Evensong in remote country churches!)

It seems to me so obvious that without some huge ground-swell of spiritual concern the transition to a more sustainable way of life remains utterly improbable. In an extraordinary dialogue between Arnold Toynbee and Daisaku Ikeda, Toynbee puts it this way:

It seems unquestionable that man's power over his environment has already reached a degree at which this power will lead to self-destruction if man continues to use it to serve his greed. People who have become addicts to greed tend to take a short-term view: 'After me, the deluge.' They may know that if they fail to restrain their greed, they will be condemning their children to destruction. They may love their children, yet this love may not move them to sacrifice part of their present affluence for the sake of safeguarding their children's future.

And he concludes this section by saying:

The present threat to mankind's survival can be removed only by a revolutionary change of heart in individual human beings. This change of heart must be inspired by religion in order to generate the will power needed for putting arduous new ideals into practice.[10]

I think I would accept this analysis, and would argue therefore that some kind of spiritual commitment, or religion in its true meaning (namely, the reconnection between each of us and the source of all life), is a fundamental part of the transformation that ecologists are talking about.

Bang goes the atheist vote! But even atheists might admit that it would be no bad thing to have a few spiritually minded allies, if only to offset some of the horrors still perpetrated in the name of religion. I can't see where the god of Ayatollah Khomeini fits into the ecological pantheon, and I can't help wondering what the connection is between the words of Christ and the rantings of Ian Paisley or the ghastly bigotry of the so-called 'Moral Majority' in the USA. A green theology will have more in common with the 'liberation theology' of many poorer countries, so that the false split between politics and religion may be removed, so that voices may be raised from every pulpit exhorting us to become 'lifeists' rather than materialists, so that we may rediscover the oneness of humanity and all creation. Thomas Merton, the American Trappist monk, puts it like this: 'We are already one. But we imagine we are not. And what we have to recover is our original unity. What we have to be is what we are.'

Only connect!

Part Four

The Green Challenge

15
Fragile Freedom

Turning Point

It should by now be apparent that green politics is not just another dimension of the disintegrating industrial world order; it is something qualitatively different. This hasn't as yet dawned on the vast majority of political commentators, and by refusing to accept the limitations of conventional left/right politics we have perhaps been asking for trouble. All political positions are open to misrepresentation, but because it is relatively new to the scene, green politics is particularly vulnerable. Distortions therefore abound, two of which are especially damaging to the long-term development of the whole movement.

The first is the notion that we are intent on taking people back rather than forwards – back to nature, back to the land, back to the bicycle even! I hope I have made it clear that in no way do ecologists contemplate a return to the primitive deprivations and discomforts of the pre-industrial age. Nor does the concept of the 'stable state' require that civilization should permanently stagnate in some sustainable stew-pond. What we are talking about is a retrieval of some of the old wisdom to inspire genuine progress through a 'civilized accommodation to nature', a state of dynamic equilibrium and harmony. Only in one respect is it correct to say that we would like to move backwards: if it is accepted that the politics of industrialism, which in chapter 4 I likened to a three-lane motorway, has indeed brought us to the edge of the abyss, then it makes a great deal of sense to take a few steps back.

The second major distortion involves the belittlement of green politics by suggesting that it relates exclusively to the physical, non-human environment. I have tried to demonstrate that the politics of radical ecology embraces *every* dimension of human experience and *all* life on Earth – that is to say, it goes a great deal further in terms of political comprehensiveness than any other political persuasion or ideology has ever gone before. As such it is the only expression of genuine opposition to the dominant world order. If there is still any doubt in your mind as to the validity of that claim, you may care to cast your eye over the full range of differences that distinguish the two world views or paradigms looked at in Parts Two and Three of this book.

The politics of industrialism	*The politics of ecology*
A deterministic view of the future	Flexibility and an emphasis on personal autonomy
An ethos of aggressive individualism	A co-operatively based, communitarian society
Materialism, pure and simple	A move towards spiritual, non-material values
Divisive, reductionist analysis	Holistic synthesis and integration
Anthropocentrism	Biocentrism
Rationality and packaged knowledge	Intuition and understanding
Outer-directed motivation	Inner-directed motivation and personal growth
Patriarchal values	Post-patriarchal, feminist values
Institutionalized violence	Non-violence
Economic growth and GNP	Sustainability and quality of life
Production for exchange and profit	Production for use
High income differentials	Low income differentials
A 'free-market' economy	Local production for local need
Ever-expanding world trade	Self-reliance

The politics of industrialism	The politics of ecology
Demand stimulation	Voluntary simplicity
Employment as a means to an end	Work as an end in itself
Capital-intensive production	Labour-intensive production
Unquestioning acceptance of the technological fix	Discriminating use and development of science and technology
Centralization, economies of scale	Decentralization, human scale
Hierarchical structure	Non-hierarchical structure
Dependence upon experts	Participative involvement
Representative democracy	Direct democracy
Emphasis on law and order	Libertarianism
Sovereignty of nation-state	Internationalism and global solidarity
Domination over nature	Harmony with nature
Environmentalism	Ecology
Environment managed as a resource	Resources regarded as strictly finite
Nuclear power	Renewable sources of energy
High energy, high consumption	Low energy, low consumption

Even the most cursory glance at such a comparison should demonstrate that the old age is giving way to the new, that the turning point is already with us. Everything that once served to enhance both individual and collective security now serves to undermine it: larger defence budget, more sophisticated weaponry, the maximization of production and consumption, higher productivity, increased GNP, the industrialization of the Third World, expanded world trade, the comprehensive exploitation of the Earth's resources, an emphasis on individualism, the triumph of materialism, the sovereignty of the nation-state, uncontrolled technological development – these were once the hallmarks of success, the guarantors of security. Collectively they now threaten our very survival. E. P. Thompson has described the determination of industrial nations to carry on down the same road as 'exterminism'. In Germany they use the word 'vitalism' to characterize the alternative.

The full extent of the crisis is now clear. The spirit of industrialism is rapidly losing its grip. The doctrine of scientific rationality and material growth has signally failed to provide people with any lasting ideals or values – so much so that the very legitimacy of the dominant world order is now in question. Duane Elgin refers to this state of limbo as an 'arrested civilization . . . one that is paralysed into dynamic inaction. The social order is expanding all of its creative energy on just maintaining the status quo. The winter of industrial civilization appears to stretch into an endless future.'[1] The writing is indisputably on the wall, but the problem is that for many the ink remains invisible.

Time is Running Out

The Chinese word for 'crisis' (*wei-chi*) means both 'danger' and 'opportunity'. Authors within the green movement are well aware of this ambivalence, yet it is surely significant that so many feel the time is now ripe to stress the opportunities rather than the dangers, be it through the kind of high-technology optimism of Alvin Toffler's *The Third Wave* or through the inner potentialities of Marilyn Ferguson's *Aquarian Conspiracy*. In *Muddling towards Frugality* Warren Johnson even manages to make a virtue of the impotence and failure of Governments to meet people's need, in as much as these should provide a stimulus to adaptation through the sort of self-reliance and co-operation that local communities must inevitably resort to. And Willis Herman draws an analogy from the world of psychotherapy to bring hope even to the most despairing breast:

> All we have learned of psychotherapy suggests that it is at the precise time when the individual feels as if his whole life is crashing down around him that he is most likely to achieve an inner reorganization constituting a quantum leap in his growth towards maturity. Our hope, our belief is that it is precisely when society's future seems so beleaguered – when its problems seem almost

staggering in complexity, when so many individuals seem alienated, and so many values seem to have deteriorated – that it is most likely to achieve a metamorphosis in society's growth towards maturity, towards more truly enhancing and fulfilling the human spirit than ever before.[2]

It may be no coincidence that all these authors are Americans. The optimism of the European greens is rather more restrained and sometimes even a little fatalistic: 'Europe will be green or not at all' is one of the more striking of the slogans of the French ecologists. We are perhaps more conscious of the extraordinary lengths to which people are prepared to go in order to avoid the implications of the green analysis, denying the evidence of their own eyes, blaming it all on a range of readily available scapegoats, indifferently pretending it doesn't really matter, or escaping into a world of hedonistic isolation. Many seem resigned to whatever may happen, and many more cling to their dependence on the powers that be.

We may also be more conscious in Europe of the habitual failure of the imagination to cope with radical change. For instance, many literally cannot think their way through to a future that is not dependent on economic growth. Common sense alone should demonstrate the benefits of working and producing things to meet people's needs *directly* rather than working and producing things to make enough money to meet people's needs *indirectly*. But the idea of any significant or long-term reduction in the material standard of living is so hard for many to take on board that politicians understandably feel under no obligation to strip down their clapped-out ideologies and start again. This, one suspects, has always been the case; the Industrial Revolution was itself carried through against the interests of the politicians and members of the aristocracy who then wielded the power in the land. The post-industrial 'revolution' will no doubt follow the same pattern: because the politicians have always got more to lose than the people whom they claim to represent, they will change their ways only when people positively oblige them to. What it all boils down to is just how much time we have to effect such a transformation.

Our real fear is not that we shall run out of oil or clean water or other vital resources, but that we shall run out of time. All societies depend upon a balance of coercion and compliance; if a Government fails to win the hearts and minds of people in persuading them to comply, then the balance inevitably swings towards coercion. And as Governments obstinately turn their backs on the future by attempting to recreate the material successes of a bygone age, we must anticipate an extended period of social turmoil. Economic hardship and personal alienation will provide the surest footing for the spread of authoritarian ideas. As Jim Garrison puts it, 'It is an axiom of history that when the people begin to question the right of their leaders to govern, the leaders question the right of the people to question.'[3] Ever more vociferous calls for a 'firm hand on the tiller' will encourage such Governments to move away from accommodation towards quasi-totalitarian measures of control, which in turn will lead us inexorably into the cycle of protest, repression, further protest, the temporary suspension of certain freedoms and unavoidable violence.

Only the most complacent of commentators could suppose that such a cycle has not already been initiated in Britain: the control of the mass media, an increase in centralized databanks, in personal surveillance and 'precautionary' telephone-tapping, the suspension of certain trade union rights, a 'restructuring' of the welfare state, a widening of wealth differentials, an isolationist, life-boat ethic, a 'survival of the fittest' philosophy, the steady military build-up, an increase in political terrorism, moves towards an armed police force and, above all, the erosion of local democracy, as seen in the rate-capping Bill and the decision taken by Mrs Thatcher's Government to do away with the Greater London Council and other metropolitan authorities in 1986: which are the causes and which the effects of a slide down such a slippery slope? Taken alone, each of these phenomena may not mean much; taken together, do they not betoken a gradual decline into a state of quasi-totalitarian democracy, in which the trappings of a parliamentary system are upheld while the power resides elsewhere?

There should be no illusions about the urgency of this: as more and more people are driven into positions of political extremism, the threat posed to democracy today is as great as the threat to the biosphere. Fascist movements are already hard at work making systematic political warfare on the rights of ethnic and other minorities. A combination of permanent recession and the indiscriminate introduction of new technologies means that many people will not only never work but will feel increasingly encouraged to endorse totalitarian solutions. When personal alienation feeds on ecological breakdown, then all we have to look forward to is a veritable 'technocracy of the ruins'.

Political parties invariably, if understandably, have difficulty in adapting themselves to conditions other than those which gave birth to them, but the cruellest irony of all is simply this: many of those who are most outspoken in their protests against the erosion of democracy are themselves deeply implicated in perpetuating the sort of politics that has inevitably brought us to such a parlous pass. Politicians of left, right and centre are all both parents and prisoners of the current crisis; and in their refusal to help liberate others through liberating themselves, they are all deeply conservative and deeply reactionary. By clinging with growing desperation to the industrial paradigm, by supposing that the politics of plenty (or what Hazel Henderson refers to as the 'politics of the last hurrah') is still the only way to achieve progress, they condemn both themselves and us. To them will fall the increasingly thankless task of dividing up a diminishing economic pie that they have promised should be getting larger; to us will fall the sordid consequences of so profound a failure collectively to get a grip on reality.

Where the Wasteland Ends

The longer we resist the inevitability of change, the less chance is there that we shall achieve it democratically; the sooner we commit ourselves to change, the easier such a process will be. Green politics has come of age just in time. Summing it all up, Lewis Mumford wrote: 'All thinking worthy of the name must now be ecological.' And being the great visionary that he was, he did not mean by 'ecology' some kind of all-purpose, reformist repair kit to patch up and protect the status quo, ascribing purely utilitarian values to the rest of creation, anthropocentric to its

inevitably bloody end; he meant ecology in all its biocentric,
holistic fullness, seeing humankind as just one strand in the
seamless web of creation, not above or outside creation but
miraculously incorporate within it. Even ecologists are only now
beginning to realize the revolutionary implications of the seeds
that they have helped to sow. Way back in 1972 Theodore
Roszak wrote a stunning book called *Where the Wasteland Ends*
(which for me personally was the first book to lift me up out of
the slough of industrial despond), in which he posed a crucial
question:

> Ecology stands at a critical cross-roads. Is it too to become another
> anthropocentric technique of efficient manipulation, a matter of
> enlightened self-interest and expert, long-range resource budget-
> ing? Or will it meet the nature mystics on their own terms, and so
> recognize that we are to embrace nature as if indeed it were a
> beloved person in whom, as in ourselves, something sacred
> dwells? The question remains open: which will ecology be, the last
> of the old sciences or the first of the new?[4]

That question is now answered, not just through the birth and
growth of green movements and green parties throughout the
Western world but also through the crucial distinction such
parties insistently draw, often to their own immediate disadvan-
tage, between 'deep ecology' and 'shallow ecology', between
genuine green politics and the sham of reformist environmentalism. In *Person/Planet* Roszak appropriately answers his own
questions:

> My purpose is to suggest that the environmental anguish of the
> Earth has entered our lives as a radical transformation of human
> identity. The needs of the planet and the needs of the person have
> become one, and together they have begun to act upon the central
> institutions of our society with a force that is profoundly subvers-
> ive, but which carries within it the promise of cultural renewal.[5]

This then is the time when we must pose the full challenge of
the green perspective. Despite genuine sympathy and profound
respect for the many 'vaguely greens' in other parties, for so-
called 'non-political' greens in a so-called 'green movement', for

progressive socialists, old-fashioned Conservatives, radical
Liberals and innocent Social Democrats, for feminists and peace
activists, for defenders of human rights and animal rights, for
followers of Christ or Gandhi or the Buddha or Lao-Tsu, and for
those who are none of the above but just love their children or
just love the Earth – I must now ask why, oh why, are you out
there directly underwriting or indirectly condoning the perpetu-
ation of soul-destroying, life-destroying industrialism?

And should I ever have the privilege of asking you that
question face to face, don't put me off with protestations about
the time not being ripe or the circumstances not propitious, for I
would then be forced to reach for the Green Party's 1983
election manifesto, *Politics for Life*, which opens with these
uncompromising words of Fritz Schumacher:

> We must do what we conceive to be the right thing and not
> bother our heads or burden our souls with whether we're going to
> be successful. Because if we don't do the right thing, we'll be
> doing the wrong thing, and we'll be part of the disease and not a
> part of the cure.

The last chapter of this book is for those whose heads are not
bothered and whose souls are not burdened by the snares of
'success' as interpreted by the present world order, and who are
prepared, in their own way, even now, to embrace the radical
alternative of green politics. Although I have addressed the
challenge primarily to those already involved in politics or in the
environment movement, it seems to me that the green alternative
makes as much sense, if not more, to those who are less readily
defined. It must, however, be said that it will be easier for some
than for others!

16
Common Ground

Whither Socialism?

Herr Kohl's favourite joke during the 1983 election campaign in West Germany was to compare die Grünen with tomatoes: they start out green and then turn red. Like so many contemporary politicians, he's got things completely the wrong way round: most of the reds who joined die Grünen in the early 1980s have since gone genuinely green. Most members of the Labour Party in Britain wouldn't even understand the joke, let alone Herr Kohl's misapprehension. An article in *The Times* in April 1984 entitled 'Let's Make the Red Flag Green', written by Robin Cook, Labour MP for Livingstone and one of Neil Kinnock's 'kitchen Cabinet', must have come as a bit of a shock to most of his colleagues who up until then had been firmly of the opinion that green politics was all about looking after the leprechauns at the bottom of the garden.

Robin Cook's article rightly referred to 'areas of powerful congruence between socialists and ecological thinking', particularly in terms of our shared commitment to achieving social justice and of our analysis of the contradictions inherent in capitalism. The consequences of higher productivity in a capital-intensive economy have, however, not yet dawned on most members of the Labour Party. As I have already demonstrated, higher productivity *without* growth in demand and output means mass unemployment; higher productivity *with* growth in demand

and output means ecological catastrophe. Capitalism can indeed survive only through permanent expansion – which in turn means the accelerating contraction of our life-support systems. But as things stand today, this is just as much of a dilemma for the Labour Party as it is for the Tories. Labour's 'Alternative Economic Strategy' demands exponential increases in economic growth of up to 7 per cent to create just $1\frac{1}{2}$ million jobs! Are we to assume that we need a 14 per cent rate of economic growth to get back to 'full employment'? This purportedly 'radical' strategy shows absolutely no interest in either the destruction of the environment or the quality of the work that might be created by such an emphasis on growth. In Robin Cook's own rather damning words: 'Neither the market capitalism of America nor the state capitalism of the Soviet Union has produced an economic model which respects the fine tolerances of nature or grants self-respect to labour.'[1]

This continuing obsession with industrial expansion, against the interests of the planet and therefore of its people, is currently reflected both in the insistence on centralized planning and an approach to Britain's industrial problems totally at odds with the views of those socialists who still hanker after the inspired decentralism of William Morris, and in the emphasis on extended nationalization programmes and the public ownership of the means of production, for which there is absolutely no case if the objectives to be pursued are as destructive and as narrow as those of privately owned big business. Those who still subscribe to an essentially Marxist analysis of the role of the state and the need to seize control of it, by force if necessary, are unlikely to be ready disciples to the green cause. It is transparently dishonest to suppose that socialists are any better at relinquishing power than those from whom they are accustomed to seize it. As James Robertson points out, if we are truly looking for a withering of the state, then we need to be thinking of a one-stage and not a two-stage process.

This also raises many problems for the trade unions. Their insistence on propping up the formal economy at all costs and enthusiastically endorsing the philosophy of industrialism (which they perceive to be the only way of defending their members' interests) means that they will find it difficult to adapt to the kind of decentralized, small-unit, co-operatively based economy that the greens are talking about. As the distinction between employer and employee are gradually eroded and working patterns become more flexible, the highly centralized, bureaucratic and often undemocratic structure of some unions will be revealed as utterly inappropriate to a new age. Rudolf Bahro, a former member of the East German Communist Party and now a leading member of die Grünen, takes the analysis even further: 'Viewed from the outside, the opposition between the trade unions and the employers' organizations is relative; the opposition of the whole set-up to the interests of humanity is absolute.'[2]

It is inevitable that greens should find themselves at odds with the conventional socialist analysis of class politics. Still to be thinking in terms of an industrial proletariat, of the masses as the vehicle of revolution and of the bourgeoisie as the implacable enemy is to miss the point about the balance of power in today's world. Among all notional 'classes' of people there are those who exercise their power responsibly in the interest of life on Earth and those who use their power to the detriment of both people and planet. A genuine redistribution of power can no longer be simplistically interpreted in terms of setting class against class, special interest against special interest: the need to serve the general interest of humanity now transcends any such old-world divisiveness. By the same analysis, the conventional socialist interpretation of equity and the distribution of wealth falls far short of what is required in an age of global interdependence. Without that reassuring cushion of economic growth, not only must we now ensure a radical redistribution in this country but we must also find viable ways of sharing our wealth with the Third World and the planet's wealth with generations as yet unborn. 'Intergenerational equity' (summed up by the statement 'We do not inherit the Earth from our ancestors; we borrow it from our children') adds a new and profoundly challenging dimension to redistributive politics – and one not likely to be met through the politics of envy and narrow sectionalism. The complex historical problem that confronts all socialists is that their philosophy and values evolved in the context of an apparently infinite physical environment: those who are still enthusiastically engaged in the exploitation of the planet, albeit for the 'best possible motives', are in the business of destroying wealth rather than redistributing it.

Such aspects of contemporary socialism are hardly surprising, given its uncompromisingly materialistic orientation, its chronic tendency to look to the white heat of technology as the means by which progress will be achieved and its ill-disguised contempt for spiritual and non-economic values. In his book *Green Politics*, Fritjof Capra records this highly significant comment of Roland Vogt, one of the twenty-seven MPs in the West German Bundestag: 'The materialist-leftish approach is destructive within the Greens. Whenever the visionary or spiritual people make a proposal, the Marxist-oriented greens neutralize it as effectively as acid.'[3] This need to strip off the materialist blinkers is closely linked in green politics to the emphasis we place on non-violence. Unlike many socialists, we do not see non-violence as just another useful tactic in the 'struggle to seize power from the enemy'. It is rapidly becoming the cornerstone of all we aspire to achieve. In the words of Martin Luther King, 'We no longer have a choice between violence and non-violence. The choice is either non-violence or non-existence.' It is no accident that much of the impetus for this shift of emphasis comes from a flourishing eco-feminist movement, and one cannot help but suppose that many within a predominantly patriarchal labour movement, despite lyrical protestations to the contrary, would find it difficult to accept the emphasis that we place on feminist values and the extent to which they underpin green politics.

These then are some of the green stripes that Robin Cook and other members of the Socialist Environment and Resources Association must imprint on their red flag. The fact that the flag would then be almost entirely green leads one to suppose that there may well be easier ways of achieving the same goal. But the movement from red to green is inevitable, for the current theory and practice of socialism is both unsustainable and unrealistic. The progressive, radical, libertarian thrust of socialism has been vitiated by its wholehearted commitment to materialist industrialism: one simply cannot cure today's problems with the means that have produced them. The task confronting 'green' socialists in

Britain is therefore enormous – and they will find that there's a
great deal more to green politics than simply nicking our slogan!

The Extremists in the Centre

When it comes to the greening of the Liberal Party, green
Liberals do not have quite such a formidable task on their hands.
It is my perception that of all the major parties in Britain, the
Liberals are the most likely, both by temperament and by reason
of political expedience, to move towards the green position. This
may seem paradoxical, given the present leadership and outlook
of the Liberal Party, but there are many who believe that this is
nothing like as fixed and unyielding as it may appear. Many of the
problems arise from the present alliance with the Social
Democratic Party (SDP), for in almost every respect the SDP
endorses and reinforces the greyest and gloomiest areas of contemporary Liberalism. The key phrase in the Alliance's 1983
general election manifesto, clearly showing David Owen's iron
fist in David Steel's woolly mitten, was simply this: 'We must
ensure Britain's economic recovery in a brutally competitive
world.' Remember Keynes and the need to pretend for another
hundred years that 'fair is foul and foul is fair: for foul is useful and
fair is not'? Remember our model industrialist exhorting those
Durham students to get out there and compete for the world's
diminishing resources 'while they're still there'? Those who
choose to live by brutal competition will assuredly die by it.

The Alliance is still fundamentally expansionary, growth-
oriented, conventionally Keynesian and reflationary in its
economics. Both parties within it still talk glibly of a return to full
employment, and both are committed to a massive increase in
world trade. Both are ardent advocates of the 'free market', and
find the idea that we may have to control the development of
new technologies utterly abhorrent. They are, in short, un-
reconstructed capitalists for the simple reason that they have not
yet realized that modern capitalism will destroy the planet long
before it manages to meet the needs of the people who depend on
that planet. All their talk about 'green growth' is therefore a
transparently obvious and dishonest ploy designed to drape a few

green trimmings over their grey old shoulders. We should not forget that in 1979 the Liberal Assembly passed a surprisingly radical 'no-growth' motion, which was promptly disowned by the Parliamentary Liberal Party and has not been heard of since. Despite all the hard work of the Liberal Ecology Group and the recent upsurge of green interest among the Young Liberals, the Liberal Party itself has actually moved back rather than forward since 1979.

It should also be pointed out that both parties are firmly 'Atlanticist' in their outlook, and the influence of the SDP has done much to delay the long overdue shift within the Liberal Party to an unequivocally unilateralist stance. Attempts by ordinary members of the party to force their leadership to disown Cruise missiles and all other American weapons based in the UK have been consistently thwarted. The Alliance manifesto for the 1984 European elections demonstrates the extent to which both parties are firmly wedded to NATO and to the special relationship with the USA. The notion of non-alignment is obviously particularly distressing to David Owen, who for many years has been a member of the Trilateral Commission, an enormously powerful group of politicians and industrialists with the specific role of promoting Western capitalism. (The Commission recently proposed that Japan, a country whose economic success has been largely dependent on the fact that it spends almost nothing on defence, should now be asked to contribute to the huge defence costs of Europe and the USA: such is the nature of 'radicalism' within such an organization.) On top of that, both parties are fervently in favour of the EEC, and both believe that the sort of proposals outlined in the Brandt Report to increase our prosperity in the process of doing something about the Third World are the only way of helping the world's poor.

Both Steel and Owen talk at one and the same time of being 'moderate' and yet 'radical'. As it happens, they are neither. The reality of contemporary centrist politics, with its inability to be anything other than an amalgam of left-wing industrialism and right-wing industrialism, means that the so-called 'moderates' of the Alliance are in fact the most extreme of all today's extremists. Their inertia, their nostalgia for the past, their partial vision, their very reasonableness and establishmentarianism ensure that even in

their most dynamic, insightful moments it is an extreme of
conservative reaction that they serve rather than any inspired
accommodation with the planet. As we stand on the threshold of
a new age, one has to be adjudged an extremist if one still
supposes that it is possible to achieve lasting security through
dependence on nuclear weapons or the build-up of conventional
arms; that it is possible to help the Third World while the First
World continues to help itself to the wealth of our finite planet; or
that it is possible to maintain present rates of economic growth
while simultaneously safeguarding the biosphere.

And it is equally absurd that the Alliance should claim to be
radical. In steadily applying itself to the alleviation of the
symptoms of industrial decline, centrist politics actually makes a
virtue of ignoring the causes or the roots of the problems. To
promote that particular blend of institutionalized reformism, to
pour the soothing balm of moderation on the so-called 'extremes'
of right and left, to recycle the tarnished successes of a bygone
age – is that really all the future offers? Without acknowledging
the primacy of an ecological analysis, the roots that any political
party puts down are by definition illusory. When looked at from
this perspective, the SDP and those Liberals who support the
Alliance are about as radical and as green as a dying elm tree.

Fortunately, many Liberals have realized the speed with which,
arm in arm with the SDP, they are disappearing down the cul-de-
sac of industrialism. This is no minority, fringe voice: the
Association of Liberal Councillors, at least two newly elected
MPs, the Young Liberals and many individual 'mainstream'
Liberals have all come to the conclusion that their party is in
danger of losing its unique purpose and political vision. Given
such a realization, there is only one direction in which the Liberal
Party can move.

Conservation or Conservatism?

Sadly, no such tendency is as yet apparent among the ranks of
modern Conservatism; the likelihood of greening such stalwarts
of industrialism as Norman Tebbit, Geoffrey Howe, Nigel
Lawson and Mrs Thatcher herself is remote indeed. And if such

people retain control of the Conservative Party, then the most probable, and perhaps the most hopeful, dialectic for the future may be seen in terms of the varying shades of green in the Green, Labour and Liberal parties taking on the varying shades of grey in the Conservative Party. (The SDP could go either way or, more likely, simply disappear.) Yet to many individual members of the Conservative Party this must seem rather surprising, for there is much about green politics that is instantly and deeply appealing to a certain kind of Tory.

Genuine concern for the environment and for the land has long been at the heart of what is now considered to be old-fashioned Conservatism. Such concerns are often combined with a deep understanding of the crucial importance of the local community and an often unconscious yet still extremely effective way of working for others through the informal, direct interrelationships that underpin any community. A profound dislike of waste, of profligacy and of the kind of misuse of both human and physical resources that characterizes our society informs many of their political attitudes, and their emphasis on spiritual, non-material values remains of considerable significance in a secular, despiritualized environment. Above all, a certain economic realism (now grievously misrepresented in the heartless vicissitudes of monetarism) should commend itself to advocates of the conserver economy. Managing the household budget *is* important – though any such housekeeping will be immeasurably enhanced if we can ensure that the house itself does not actually fall down around us. Not even the Conservative Party can ignore for much longer the logic of ecology.

It would therefore be foolish and blinkered for greens to suppose that merely because a person thinks and lives conservatively, he or she is incapable of embracing a green alternative. However, it would be equally foolish to suppose that the present manifestation of Conservatism is anything but disastrous both for people in Britain and for the planet itself. This particular Tory Government falls even at the first hurdle: one wouldn't expect it to understand anything about green politics, but it might at least be doing something about the environment. In its 1983 general election manifesto it had the impertinence to claim that 'No Government has done more for the environment.' Such mendacity must be deeply shameful to many members of the Conservative Party, especially if one thinks of this Government's efforts to hold up action on acid rain, its decision to get rid of the Waste Management Advisory Council, its long delay in implementing crucial sections of the 1974 Control of Pollution Act, its contempt for the recommendations of the Royal Commission on Environmental Pollution, its cutbacks in the numbers of environmental health officers and factory inspectors, its neglect of water and sewage systems, its threat to the Green Belt, its delaying tactics as regards the introduction of Environmental Impact Statements, its refusal to amend the disastrous Wildlife and Countryside Act – I could go on for pages. The simple fact is that by doing so little, no Government has ever done more to damage the environment.

All of which presents a peculiar challenge to an environment movement which has long prided itself on political neutrality and even-handedness in its dealings with mainstream political parties. I do not believe that such an approach is any longer viable. It is time that the environment movement acknowledged just how important a role it has to play in terms of promoting social and economic change. Even if it is to win the many vital single-issue campaigns in which it is involved, it must cease to see itself as a bunch of nice people for ever waiting in the wings and cease to operate as if it were peripheral to mainstream political concerns. The environment movement must move resolutely centre-stage, which will inevitably involve a more pronounced confrontation with those who presently monopolize that position. And it must at the same time develop the skills to communicate the immediacy and the universality of environmental concern to people in every walk of life, not just to the well read and the highly motivated.

This will require a far more rigorous analysis of the *causes* of the environmental crisis and a far more open advocacy of the sorts of structural change that will be necessary to effect any real improvements. Concern for the environment provides as good a starting point as any for green politics. But unless it then encompasses fundamental social and economic issues, it will have contributed little towards eliminating the root causes of that crisis. If it stops at mere reforms in conservation and pollution control, then it will be operating simply as a leaky safety valve for the existing systems of exploitative politics. The sort of environmental engineering we see so much of today (and by virtue of which many politicians clumsily lay claim to some kind of mottled green tinge!) only serves to obscure the real problems.

The political challenge to the environment movement is simply this: it is impossible to end the exploitation of the environment without bringing to an end the exploitation of our fellow human beings. In his book *Ecology and Social Action*, Barry Commoner wrote:

> When any environmental issue is probed to its origins, it reveals an inescapable truth – that the root cause of the crisis is not to be found in how men interact with nature, but in how they interact with each other; that to solve the environmental crisis we must solve the problem of poverty, racial injustice and war; that the debt to nature, which is the measure of the environmental crisis, cannot be paid, person by person, in recycled bottles or ecologically sound habits, but in the ancient coin of social justice.[4]

Politics for Life

Yet even that is not enough. To the enduring challenge of social justice we must now add the challenge of spiritual enlightenment. In the same way that environmentalists can no longer deny the radical implications of their commitment, so people involved today in religious and spiritual concerns must appreciate that political action has of necessity to be part of those concerns. The challenge of Christ, of Gandhi, of all great spiritual leaders, has always been as much political as spiritual. From a green perspective it works both ways, interdependently, indivisibly:

> With the holistic sense of spirituality, one's personal life is truly political and one's political life is truly personal. Anyone who does not comprehend within him- or herself this essential unity cannot achieve political change on a deep level and cannot strive for the ideals of the greens.[5]

Petra Kelly's words do indeed work at a level far deeper than that which contemporary industrial politics is either willing or able to handle. Stripped of a spiritual dimension, politics in today's world is a hollow shell, and religion stripped of its political dimension is irresponsibly escapist. There is *no* place today where we can escape to, no sanctuary of the soul, no island hide-out, no inner or outer refuge that can prevent us from experiencing the plight of the world and all its people. To suppose that among all this one might remain neutral or disengaged represents the final triumph of industrial alienation. As Paulo Freire says, 'Washing one's hands of the conflict between the powerful and the powerless means to side with the powerful, not to be neutral.' And power, as exercised today, drags us all inexorably towards the abyss.

So many challenges! I sometimes think that I would never have got involved in green politics had I really understood what it was that I was getting involved in. And I am sometimes apprehensive that what we now seem to be asking of people may appear to be so intimidating that it's easier for them not even to take the first step. I profoundly hope that such has not been the overall effect of this book, for when it comes right down to it, all I am actually asking of people is that they should consider four questions and then take action on the basis of their own answers.

First, how much evidence does each of us require before we realize that the politics of industrialism has irretrievably lost its way and lost its soul, and that today's winners can claim their prizes only at the expense of tomorrow's losers, namely, the vast majority of humankind, the Earth itself and those unborn generations that will have to pick up the pieces?

Secondly, given such evidence, how long will it take us to move from the politics of negation to the politics of affirmation, to move beyond what Marcuse called the 'great refusal', to realize that saying no to nuclear weapons is not enough if we wish to establish the conditions for lasting peace, that saying no to the worst excesses of industrialism is not enough if we want to help today's poor, unemployed and disadvantaged?

Thirdly, given the readiness to make such an affirmation, when will the few become the many in realizing that the *only* alternative to the politics of exploitation and class interests is the politics of ecology and life interests, and that we must affirm such an approach not just in the way we vote but also in our way of life, in our relationships with each other and in our moral and spiritual beliefs?

And, last, given such awareness, when will we find the courage to do *now* what is easier to put off until tomorrow, to accept unequivocally that each of us weaves our own strand in the web of life, and that in the power we have to transform our own lives we have also the power to transform life on Earth?

Each of us will have different answers to these questions. Though I have referred at the end of this book to a few books to be read or organizations to be joined, the green model of social and political change insists that each of us should find our own way of living and seeing green. We should not be waiting around for cataclysmic warnings or charismatic leaders: what matters is that we should set out now or push on further down the right road, establishing our common ground as we go, developing our 'common sense' of what it means to be working together for a better future.

To avoid writing the Earth's obituary we must cease to see the future simply as an extension of the present, and we must think as much about what *should be* as about what actually *is*. We must think again of the links between ourselves and the Earth, and of the way the Earth speaks to us through an ideal of life. We must seek ways creatively to disintegrate the economic and industrial constraints that are turning our world and our lives into a wasteland. Above all, we must learn to blend our concern for people with our respect for the Earth through the post-industrial politics of peace, liberation and ecology: the politics of life.

References

1   **Green Politics Today**

    1   Aurelio Peccei, *100 Pages for the Future*, London, Futura Books, 1982, p. 132.
    2   Michael Allaby, *The Eco-Activists: Youth Fights for a Human Environment*, Croydon, Charles Knight, 1971.
    3   Theodore Roszak, *Person/Planet*, St Albans, Granada, 1981.

2   **Opposing World Views**

    1   Ralf Dahrendorf, *After Social Democracy*, London, Liberal Party Publications, 1981.

3   **For Earthly Reason**

    1   *The Global 2000 Report to the President*, Harmondsworth, Penguin, 1981, p. 1.
    2   Teresa Hayter, *The Creation of World Poverty*, London, Pluto Press, 1981, p. 25.
    3   Eric McGraw, *Proposals for a National Policy on Population*, London, Population Concern, 1981, p. 15.
    4   UN Food and Agriculture Organization, *Soil and Water Conservation*, Rome, 1980.
    5   Erik Eckholm, *Down to Earth*, London, Pluto Press, 1982, p. 141.
    6   Lester Brown, *Building a Sustainable Society*, New York, W. W. Norton, 1981, p. 13.
    7   Eckholm, *Down to Earth*, p. 161.
    8   Fritz Schumacher, *Small is Beautiful*, Tunbridge Wells, Abacus, 1974, p. 98.
    9   Friends of the Earth, *Pesticides: The Case of an Industry out of Control*, London, 1984.

10 Organization for Economic Co-operation and Development, *Environment Policies for the Eighties*, Paris, 1980, p. 12.
11 Schumacher, *Small is Beautiful*, p. 11.

4   Industrialism in All its Glory

1   Roszak, *Person/Planet*.
2   Narindar Singh, *Economics and the Crisis of Ecology*, Delhi, Oxford University Press, 1976.
3   Ibid.
4   Lewis Mumford, *The Pentagon of Power*, London, Secker & Warburg, 1970, p. 185.

5   The World at War

1   Edward Thompson, *Beyond the Cold War*, London, Merlin Press, 1982, p. 17.
2   *The Church and the Bomb*, London, Hodder & Stoughton, 1982.
3   Lord Zuckerman, *Science Advisers, Scientific Advisers and Nuclear Weapons*, London, Menard, 1980, p. 10.
4   Robert Scheer, *With Enough Shovels*, London, Secker & Warburg, 1983.
5   Jim Garrison, *The Russian Threat*, London, Gateway Books, 1983, p. 304.
6   *New Internationalist*, March 1983.

6   The Collapse of Economics

1   Kathleen Newland, *Global Employment and Economic Justice*, Washington, DC, Worldwatch Paper 28, 1979.
2   Hazel Henderson, *Creating Alternative Futures*, New York, Berkley Windhover, 1978.
3   Lester Brown, *Population Policies for a New Economic Era*, Washington DC, Worldwatch Paper 53, 1983.
4   Robert Fuller, *Inflation: The Rising Cost of Living on a Small Planet*, Washington DC, Worldwatch Paper 34, 1980.
5   Brown, *Building a Sustainable Society*, p. 120.
6   Colin Sweet, *The Costs of Nuclear Power*, Sheffield, Anti-Nuclear Campaign, 1982, pp. 39–40.
7   Duncan Burn, Select Committee Report, vol. 4, pp. 1253–4.

7   Alienation is a Way of Life

1   Stephen Cotgrove, *Catastrophe or Cornucopia*, Chichester, John Wiley & Sons, 1982, p. 82.

2 Campaign for Freedom of Information, *Protecting the Polluter*, Secrets File No. 2, 1984.

8   A System without a Soul

1 Schumacher, *Small is Beautiful*, p. 26.
2 J. M. Keynes, 'Economic Possibilities for our Grandchildren', in *Essays in Persuasion*, London, Macmillan, 1931.
3 David Ehrenfeld, 'The Conservation of Non-Resources', *American Scientist*, November 1976.
4 Friends of the Earth, *Proposals for a Natural Heritage Bill*, London, 1983.
5 Robert Waller, *The Agricultural Balance Sheet*, London, Green Alliance, 1980, p. 3.
6 Laurens van der Post, *Jung and the Story of Our Time*, London, Hogarth Press, 1976.
7 Henryk Skolimowski, *Economics Today: What Do We Need?*, London, Green Alliance, 1980, p. 10.
8 Schumacher, *Small is Beautiful*.
9 Roszak, *Person/Planet*, p. 15.
10 Ibid., p. 26.

9   Ecologics

1 Cotgrove, *Catastrophe or Cornucopia*, p. 48.
2 Duane Elgin, *Voluntary Simplicity*, New York, William Morrow, 1981, p. 199.
3 Ivan Illich, *Energy and Equity*, London, Calder & Boyars, 1973, p. 30.
4 R. H. Tawney, *Religion and the Rise of Capitalism*, Harmondsworth, Penguin, 1948.
5 Schumacher, *Small is Beautiful*, p. 33.

10   Green Economics

1 Ecology Party, *Working for a Future*, London, 1980.
2 Mumford, *The Pentagon of Power*, p. 153.
3 Singh, *Economics and the Crisis of Ecology*.
4 Joseph Kraft, 'A Talk with Trudeau', *Washington Post*, 17 May 1977.

11   Green Peace

1 Garrison, *The Russian Threat*, p. 320.

2 Jonathan Schell, *The Fate of the Earth*, New York, Alfred A. Knopf, 1982, p. 153.

3 Martin H. Ryle, *The Politics of Nuclear Disarmament*, London, Pluto Press, 1981, p. 14.

4 Lord Carver, 'What the Bloody Hell is it for?', *Sanity*, December 1982.

5 *North–South: A Programme for Survival*, London, Pan, 1980.

6 Norman Myers, *Guardian*, 3 June 1982.

7 Garrison, *The Russian Threat*, p. 323.

12   Sustainable Society

1 Schumacher, *Small is Beautiful*, p. 54.

2 James Robertson, *The Sane Alternative*, London, published privately, 1983, p. 71.

3 Fritjof Capra, *The Turning Point*, London, Wildwood House, 1982, p. 365.

4 Sweet, *The Costs of Nuclear Power*, p. 31.

5 Gerald Leach *et al.*, *A Low Energy Strategy for the United Kingdom*, London, IIED, 1979, p. 259.

6 Christopher Flavin, *Energy and Architecture: The Solar Conservation Potential*, Washington, DC, Worldwatch Paper 40, 1980.

7 Leach *et al.*, *A Low Energy Strategy for the United Kingdom*.

13   A Green and Pleasant Land

1 Eckholm, *Down to Earth*, p. 195.

2 Peccei, *100 Pages for the Future*, p. 108.

14   Spirits of the Future

1 Elgin, *Voluntary Simplicity*, p. 160.

2 R. H. Tawney, *The Acquisitive Society*, Harmondsworth, Penguin, 1921.

3 Brown, *Building a Sustainable Society*, p. 349.

4 Tom Bender, 'Why We Need to Get Poor Quick', *Futurist*, August 1977.

5 Skolimowski, *Economics Today: What Do We Need?*, p. 15.

6 Simple Living Collective of San Francisco, *Taking Charge*, New York, Bantam, 1977.

7 Barry Commoner, *The Closing Circle*, New York, Alfred A. Knopf, 1971, p. 292.

8 Capra, *The Turning Point*, p. 18.

9 Erik Dammann, *The Future in Our Hands*, Oxford, Pergamon
     Press, 1979, p. 167.
10 *The Toynbee–Ikeda Dialogue*, Tokyo, Kodansha, 1976, p. 41.

15 Fragile Freedom

1 Elgin, *Voluntary Simplicity*, p. 110.
2 Willis Harman, *An Incomplete Guide to the Future*, New York,
     Norton, 1979.
3 Garrison, *The Russian Threat*, p. 235.
4 Theodore Roszak, *Where the Wasteland Ends*, London, Faber &
     Faber, 1972, p. 403.
5 Roszak, *Person/Planet*, p. 15.

16 Common Ground

1 Robin Cook, 'Towards an Alternative Ecological Strategy', *New
     Ground Magazine*, No. 2, Spring 1984.
2 Rudolf Bahro, *From Red to Green*, London, Verso, 1984.
3 Fritjof Capra, *Green Politics*, New York, E. P. Dutton, 1984,
     p. 27.
4 Barry Commoner, *Ecology and Social Action*, Berkeley, Univer-
     sity of California Press, 1973, p. 21.
5 Capra, *Green Politics*, p. 55.

Select Bibliography

Recommending books to anyone else is a remarkably risky business, so I've simply settled on the twenty-five books (arranged alphabetically by author) that have meant the most to me personally over the last five years, in the hope that such a list may be useful to others going down the same road.

The Alternative Defence Commission, *Defence Without the Bomb*, London, Taylor & Francis, 1983
 If you're prepared to work your way, step by step, through the whole non-nuclear defence case, this is the best possible book. It is authoritative in its research, cogent in all its considerations, and extremely convincing in its conclusions.

Rudolf Bahro, *From Red to Green*, London, Verso, 1984
 Bahro is the most influential and most interesting of the many socialists who now see green rather than red. The interviews that make up the book are an honest and fascinating insight into that complicated transition.

Murray Bookchin, *Toward an Ecological Society*, Montreal, Black Rose Books, 1980
 Those with any sympathy for the anarchist antecedents of the green perspective will find Bookchin's writing absorbing and highly entertaining; those who assume no such sympathy should allow Bookchin's punchy prose to show them the error of their ways.

Lester Brown, *Building a Sustainable Society*, New York, W. W. Norton, 1981
 This is an extraordinary synthesis of the best of the Worldwatch Institute; though a little short on the political implications, its insight and

authority provide an unanswerable case for a move towards sustainability.

Fritjof Capra, *The Turning Point*, London, Wildwood House, 1982
   This is a brilliant book. Give yourself plenty of time for it, for it goes deep down into the philosophical roots of our crisis before bringing you gently up into some of the answers. It was certainly a turning point for me.

Fritjof Capra and Charlene Spretnak, *Green Politics*, London, Hutchinson, 1984
   This is something of an insider's book for through-and-through greens, but its account of the history and present state of the West German greens is of considerable general interest.

Herman Daly (ed.), *Toward a Steady-State Economy*, San Francisco, W. H. Freeman, 1973
   It may be more than ten years old, but this book still bubbles with originality. Most books on green economics take this collection of essays as one of their fundamental starting points.

Erik Dammann, *The Future in Our Hands*, Oxford, Pergamon Press, 1979
   Very much a book for those who have come to believe that the problems of the Third World are somehow beyond our reach, this is a powerful and passionate challenge to each of us to amend our own lifestyle.

Erik Eckholm, *Down to Earth*, London, Pluto Press, 1982
   A systematic and quite compelling analysis of the problems of the Third World, reconciling development and ecology as no other book has managed to do since Barbara Ward's early trail-blazer, *Only One Earth*.

Duane Elgin, *Voluntary Simplicity*, New York, William Morrow, 1981
   This is a lovely book, wandering quite freely from historical angles to futuristic conjecture and all the while telling us, simply and clearly, how we can all become genuinely wealthy.

Erich Fromm, *To Have or To Be*, London, Abacus, 1979
   That acquisitive materialism is not the only option for humankind is a view shared by many; Fromm's simple yet eloquent account of the alternative makes a deep and lasting impression.

Jim Garrison, *The Russian Threat*, London, Gateway Books, 1983
I wish this book was required reading for all politicians, civil servants,
journalists and teachers, for it works at many different levels, and on all
of them Garrison shows us where the real path to peace is to be found.

Hazel Henderson, *The Politics of the Solar Age*, New York, Anchor/
Doubleday, 1981
This, and the equally impressive *Creating Alternative Futures* (New
York, Berkley Windhover, 1978), provide an excellent way into green
economics. Both books sparkle with new ideas and new insights – and
succeed in taking conventional economics irreparably to pieces!

Barry Jones, *Sleepers, Wake!*, Brighton, Wheatsheaf Books, 1982
This is the best contribution on work and technology from any
conventional politician – albeit an Australian one. It is lucid, provoca-
tive, extremely constructive and just about radical enough to hit the
mark.

Jim Lovelock, *Gaia: A New Look at Life on Earth*, Oxford, Oxford
University Press, 1979
This is a wonderfully inspirational book about what makes life
possible on our planet, and Lovelock's direct and sympathetic style
makes it accessible to even the least scientifically minded.

Aurelio Peccei, *100 Pages for the Future*, London, Futura, 1982
Peccei has put together a pretty nifty little summary of why we are
about to go over the edge of the abyss and what we should be doing to
stop ourselves. A little short on political reality but a good introductory
handbook.

Kit Pedler, *Quest for Gaia*, St Albans, Granada, 1979
If you want to settle down and sort out what you have got to do to
make it all work, this is the book. From hand-made soap to simple
thermodynamics, Pedler gives us the joyful, human face of sustainability.

James Robertson, *The Sane Alternative*, published privately, 1978
This was something of a breakthrough in its time, and it still provides
a masterly, though emphatically non-political, account of the different
options before us and why there is really only one possible choice.

Theodore Roszak, *Person/Planet*, St Albans, Granada, 1981
If I had to name the one book that has made the greatest impact on
me, this would be it. It may be a bit of a mind-blower for those easing
their way in, but don't put it off for long!

Kirkpatrick Sale, *Human Scale*, London, Secker & Warburg, 1980
A monster book about getting the scale of things right – but well worth the time it takes to get through it, if only for its effective demonstration that decentralization is both viable and enlightening!

Jonathan Schell, *The Fate of the Earth*, New York, Alfred A. Knopf, 1982
Schell gets closer to explaining the suicidal impulses of the nuclear age than any other author, yet despite its comprehensive analysis of the effect of a nuclear exchange on the biosphere, it is still an extraordinarily hopeful book.

Fritz Schumacher, *Small is Beautiful*, London, Abacus, 1974
Veritably a golden oldie, yet still far more radical than many who now lay claim to Schumacher's ideas can even begin to cope with. A milestone well worth revisiting – and one from which many more will no doubt start out on the green road.

Narindar Singh, *Economics and the Crisis of Ecology*, Delhi, Oxford University Press, 1976
For those who like it strong and polemical, this is the most powerful indictment of industrialism I have ever read. A committed socialist, Singh none the less offers just as much of a challenge to the left as to the right and centre.

Henryk Skolimowski, *Eco-Philosophy*, London, Marion Boyars, 1981
By far the best book when it comes to explaining why industrialism is philosophically corrupt, barren and suicidal as well as everything else – an excellent introduction to this crucial aspect of the green perspective.

Des Wilson (ed.), *The Environmental Crisis*, London, Heinemann, 1984
This is the most comprehensive and up-to-date account of what is going on in the UK right now; if individual Friends of the Earth need further stimulus to get themselves truly radicalized, this should surely provide it!

Publications of the Green Party

Politics for Life (the Ecology Party's 1983 General Election manifesto)
Jobs for Keeps (employment pamphlet)
Working for a Future (a booklet about green economics)
Towards a Green Europe (Common Programme for Action of the European Green Parties, 1984)

Also available from the Green Party's office:
Embrace the Earth: A Green View of Peace (a Green CND pamphlet)
Programme of the German Green Party (the 1983 election manifesto)

Available from the Green Alliance,
60 Chandos Place, LONDON WC2N 4HG

Maurice Ash, *Green Politics: The New Paradigm*
John Lane, *The Death and Resurrection of the Arts*
Henryk Skolimowski, *Economics Today: What Do We Need?*
Robert Waller, *The Agricultural Balance Sheet*

This book is not printed on recycled paper. This is not for lack of trying, and is more upsetting to the author than to any of his readers. It is just that the irrationality of contemporary economics makes such a use of recycled paper prohibitively expensive.

Green Organizations

UK

Ecology Party
36–38 Clapham Road, London
SW9 0JQ

Friends of the Earth
377 City Road, London
EC1V 1NA

Greenpeace
36 Graham Street, London N1 8LL

Women for Life on Earth
2 St Edmund's Cottages,
Bove Town, Glastonbury,
Somerset BA6 8JD

Liberal Ecology Group
28 Sims Close, Romford, Essex

Socialist Environment and
Resources Association (SERA)
Poland Street, London
W1V 3DG

Australia

Friends of the Earth
366 Smith Street, Collingwood,
Victoria 3066

The Australian Democrats
P.O. Box 20905, GPO Melbourne,
Victoria 3001

The Wilderness Society
c/o Bob Brown,
Parliament House, Hobart,
Tasmania

New Zealand

Friends of the Earth
PO Box 39065, Auckland

The Values Party
PO Box 814, Auckland
PO Box 137, Wellington
PO Box 4393, Christchurch

Index